Innovation: A Very Short Introduction

'Despite the differences in surname, Mark Dodgson and I are brothers. I have known him and his faults all his life. How he wrote a book like this with David Gann I have no idea, but here it is, and a very good book too.

It tells a fascinating story, and one of growing importance. The ability to innovate is both expected and valued in the worlds of science and the arts: here we read about its importance in the field of business, and about how vastly our lives have changed— and continue to change—because of the innovation talents of individuals, and the innovation strategies of forward-thinking companies. There is a great deal here to fascinate not only those who are professionally engaged in business, but everyone who takes an intelligent interest in how the world is managed.'

Philip Pullman

'Innovation driven by human ingenuity is key to creating a more sustainable and equitable future for all.'

Paul Polman, CEO, Unilever

VERY SHORT INTRODUCTIONS are for anyone wanting a stimulating and accessible way into a new subject. They are written by experts, and have been translated into more than 45 different languages.

The series began in 1995, and now covers a wide variety of topics in every discipline. The VSI library currently contains over 550 volumes—a Very Short Introduction to everything from Psychology and Philosophy of Science to American History and Relativity—and continues to grow in every subject area.

Very Short Introductions available now:

ACCOUNTING Christopher Nobes
ADOLESCENCE Peter K. Smith
ADVERTISING Winston Fletcher
AFRICAN AMERICAN RELIGION
    Eddie S. Glaude Jr
AFRICAN HISTORY John Parker and
    Richard Rathbone
AFRICAN RELIGIONS
    Jacob K. Olupona
AGEING Nancy A. Pachana
AGNOSTICISM Robin Le Poidevin
AGRICULTURE Paul Brassley and
    Richard Soffe
ALEXANDER THE GREAT
    Hugh Bowden
ALGEBRA Peter M. Higgins
AMERICAN HISTORY Paul S. Boyer
AMERICAN IMMIGRATION
    David A. Gerber
AMERICAN LEGAL HISTORY
    G. Edward White
AMERICAN POLITICAL HISTORY
    Donald Critchlow
AMERICAN POLITICAL PARTIES
    AND ELECTIONS
    L. Sandy Maisel
AMERICAN POLITICS
    Richard M. Valelly
THE AMERICAN PRESIDENCY
    Charles O. Jones
THE AMERICAN REVOLUTION
    Robert J. Allison
AMERICAN SLAVERY
    Heather Andrea Williams
THE AMERICAN WEST Stephen Aron

AMERICAN WOMEN'S HISTORY
    Susan Ware
ANAESTHESIA Aidan O'Donnell
ANALYTIC PHILOSOPHY
    Michael Beaney
ANARCHISM Colin Ward
ANCIENT ASSYRIA Karen Radner
ANCIENT EGYPT Ian Shaw
ANCIENT EGYPTIAN ART AND
    ARCHITECTURE Christina Riggs
ANCIENT GREECE Paul Cartledge
THE ANCIENT NEAR EAST
    Amanda H. Podany
ANCIENT PHILOSOPHY Julia Annas
ANCIENT WARFARE
    Harry Sidebottom
ANGELS David Albert Jones
ANGLICANISM Mark Chapman
THE ANGLO-SAXON AGE John Blair
ANIMAL BEHAVIOUR
    Tristram D. Wyatt
THE ANIMAL KINGDOM
    Peter Holland
ANIMAL RIGHTS David DeGrazia
THE ANTARCTIC Klaus Dodds
ANTHROPOCENE Erle C. Ellis
ANTISEMITISM Steven Beller
ANXIETY Daniel Freeman and
    Jason Freeman
APPLIED MATHEMATICS
    Alain Goriely
THE APOCRYPHAL GOSPELS
    Paul Foster
ARCHAEOLOGY Paul Bahn
ARCHITECTURE Andrew Ballantyne

Available soon:

For more information visit our website

www.oup.com/vsi/

Mark Dodgson and David Gann

INNOVATION

A Very Short Introduction

SECOND EDITION

OXFORD
UNIVERSITY PRESS

OXFORD

UNIVERSITY PRESS

Great Clarendon Street, Oxford, OX2 6DP,
United Kingdom

Oxford University Press is a department of the University of Oxford.
It furthers the University's objective of excellence in research, scholarship,
and education by publishing worldwide. Oxford is a registered trade mark of
Oxford University Press in the UK and in certain other countries

First edition published 2010
Second edition published 2018

Impression: 1

Published in the United States of America by Oxford University Press
198 Madison Avenue, New York, NY 10016, United States of America

British Library Cataloguing in Publication Data

Data available

Library of Congress Control Number: 2018939296

ISBN 978-0-19-882504-3

Printed in Great Britain by
Ashford Colour Press Ltd, Gosport, Hampshire

For Sheridan and Anne

Contents

Preface

When we were born, not so very long ago, there were no information technologies or television companies, and airline travel was rare and luxurious. Our parents were born into a world even more different than today's, where television had yet to be invented, and there was no penicillin or frozen food. When our grandparents were born, there were no internal combustion engines, aeroplanes, cinemas, or radios. Our great grandparents lived in a world with no light bulbs, cars, telephones, bicycles, refrigerators, or typewriters, and their lives probably had more in common with a Roman peasant than with ours. In the relatively short period of 150 years, our lives at home and work have been completely transformed by new products and services. The reason why the world has changed so much can be explained in large part by innovation.

This *Very Short Introduction* defines innovation as ideas, successfully applied, and explains why it has the ability to affect us so profoundly. It will describe how innovation occurs; what and who stimulates it; how it is pursued and organized; and what its outcomes are, both positive and negative. It will argue that innovation is essential to social and economic progress, and yet that it is hugely challenging and bedevilled with failure. It describes how innovation has many contributors and takes different forms, adding to its complexity. It provides an analysis of the innovation process; the ways organizations marshal their resources to innovate; and the eventual outcomes of innovation, which can take a number of forms.

Innovations are found not only in the activities organizations do, but how they do them. The innovation process is presently going through a period of change, stimulated in large part by the opportunities of using new digital technologies. The potential sources of innovation are growing rapidly. There are, for example, more scientists and engineers alive today than in past history combined. Furthermore, the locus of innovation is changing as economies become dominated by service sectors and the ownership of, or access to, knowledge is ever more valuable compared to physical assets. Innovation is becoming more internationalized, with important new sources emerging in China, India, and elsewhere outside of the industrial powers of Europe, North America, and Japan. We explore the extent to which our understanding of innovation, developed over the past century or more, might be applied to deal with the restless transformations and turbulence we will witness in the global economy in the future.

The first three chapters explain what innovation is, its importance, and its outcomes. The subsequent chapters examine the contributors to innovation and how it is organized, and speculate on its future.

Our understanding of innovation is based on our research into countless innovative organizations around the world and our learning from the accumulated efforts of numerous scholars in the international innovation research community. Our grateful thanks are extended to all those innovators, and students of innovation, who make our journey so exciting and rewarding. We especially acknowledge Irving Wladawsky-Berger and Gerard Fairtlough, two great innovators who have had profound influences on our thinking.

Innovation is about change, and this second edition examines issues that are stimulating interest now and were non-existent or under-appreciated when we wrote the first edition. Some changes have happened very rapidly: smart phones had only just appeared in 2010, and it is estimated they will have three billion users in 2020. Some of the new technologies we describe have profoundly affected innovation in a very short period of time. Aspects of machine learning existed only in the realm of science fiction until recently, and the avalanche in the availability of data fuels much contemporary innovation. The speed of development of companies in areas such as the shared economy and electric cars has been unprecedented. In the field of ecommerce, Amazon has progressed in twenty years from a start-up to a company with over $100 billion annual sales. The enthusiasm for innovation in countries such as China and India was foreseen, but its sheer scale and the way it has become so embedded in national policies, organizational strategies, and personal ambitions is quite new. These changes feed our intention to inform the reader about the best insights from the past, the contemporary nature of innovation, and to speculate on what it will look like in the future.

List of illustrations

Chapter 1
Josiah Wedgwood: the world's greatest innovator

We begin with a study of an exemplary innovator, a person who tells us a great deal about the innovator's agenda. He established an enduring, high-profile company creating substantial innovations in the products made, the ways they were produced, and the manner in which they created value for himself and his customers. He made significant contributions to building national infrastructure, helped create a dynamic regional industry, pioneered new export markets, and positively influenced government policies. His outstanding scientific contribution was recognized by election as a Fellow of the Royal Society. He was a marketing genius, and bridged the scientific and artistic communities by a wholly new approach to industrial design. His most important contribution lay in the way he improved the quality of life and work in the society in which he lived. He is the potter Josiah Wedgwood (1730–95; Figure 1).

Born in modest circumstances into a family of Staffordshire potters, Wedgwood was the youngest of thirteen children, and his father died when he was young. He was put to work as a potter when he was 11. He suffered badly from smallpox as a child and this had a big impact on his life. As William Gladstone put it, his disease 'sent his mind inwards, it drove him to meditate upon the laws and secrets of his art...and made for him...an oracle of his own inquiring, searching, meditative, fruitful mind'. For the first

1. Josiah Wedgwood, the world's greatest innovator.

part of his career, he worked in a number of partnerships, studying every branch of the manufacture and sale of pottery. By the time Wedgwood began his own business, aged 29, he had mastered every aspect of the pottery industry.

In his mid-30s, the lameness resulting from smallpox proved too much of a constraint, so he had his leg amputated, without, of course, the aid of antiseptic or anaesthetic. As testament to his energy and drive, he was writing letters within a couple of days. A few weeks later, he suffered the tragic loss of one of his children, but he was back at work within a month of the funeral.

By the mid-18th century, the European ceramics industry had been dominated by Chinese imports for around 200 years. Chinese porcelain, invented nearly 1,000 years before, achieved a

quality in material and glaze that could not be matched.
It was much prized by the wealthy, but was too expensive for
the expanding industrial classes whose incomes and aspirations
were growing during this period of the Industrial Revolution.
Trade restrictions on Chinese manufactures further increased the
price of imports into Britain. The situation was ripe for innovation
to provide attractive, affordable ceramics for a mass market.

Wedgwood was a product innovator, constantly searching for
novelty in the materials he used, and in glazes, colours, and
design forms of his wares. He applied extensive trial-and-error
experiments to continually improve quality by removing
impurities and making results more predictable. His favourite
motto was 'Everything yields to experiment'. Some innovations
resulted from incremental improvements to existing products.
He refined a new cream-coloured earthenware being developed
in the industry at the time, transforming it into a high-quality
ceramic that was very versatile in that it could be thrown on a
wheel, turned on a lathe, or cast. After producing a dinner service
for Queen Charlotte, wife of George III, and receiving her
approval, he named this innovation 'Queen's Ware'. Other
innovations were more radical. In 1775, after around 5,000
recorded experiments that were often difficult and expensive, he
produced Jasper, a fine ceramic, commonly blue in colour. This
was one of the most significant innovations since the invention
of porcelain. His major product innovations were still being
produced by the Wedgwood company more than 200 years later.

He collaborated with numerous artists and architects in the
design of his products, including George Hepplewhite, the
furniture maker; Robert Adam, the architect; and George Stubbs,
the artist. One of his great achievements was the application of
design to the everyday. The renowned sculptor John Flaxman,
for example, produced inkstands, candlesticks, seals, cups,
and teapots. Products that were previously unattractive were
made elegant.

Wedgwood searched everywhere for ideas for designs, from customers, friends, and rivals. He looked in museums and great houses, and trawled antique shops. One valuable source of designs was a coterie of amateur artists among well-bred women. Part of Wedgwood's successful approach to working with artists, according to Llewellyn Jewitt, his 19th-century biographer, lay in his effort 'to sharpen the fancy and skill of the artist by a collision with the talents of others'.

In a speech by William Gladstone, a generation after Wedgwood's death, he says of the potter:

> His most signal and characteristic merit lay...in the firmness and fullness of his perception of the true law of what we term industrial art, or in other words, the application of the higher art to industry: the law which teaches us to aim first at giving to every object the greatest possible degree of fitness and convenience for its purpose, and next making it the vehicle for the highest degree of beauty, which compatibility with the fitness and convenience it will bear: which does not substitute the secondary for the primary end, but recognises as part of the business the study to harmonize the two.

In his manufacturing process innovations, Wedgwood introduced steam power into his factory, and as a result the Staffordshire pottery industry was the earliest adopter of this new technology. Steam power brought many changes to production processes. Previously the potteries were distant from the mills that provided power for mixing and grinding raw materials. Having power on-site significantly reduced transportation costs. It also mechanized the processes of throwing and turning pots, previously driven by foot or hand wheels. Technology enhanced efficiency in the way the use of lathes to trim, flute, and checker products improved production throughput.

He was preoccupied with quality, and spent vast amounts on pulling down and rebuilding kilns to improve their performance. Famously intolerant of poor product quality, legend has him prowling the factory smashing substandard pots and writing in chalk 'this won't do for Josiah Wedgwood' on offending workbenches.

One of the perennial challenges of making ceramics was measuring high temperatures in kilns in order to control the production process. Wedgwood invented a pyrometer, or thermometer, that recorded these temperatures, and for this achievement he was elected a Fellow of the Royal Society in 1783.

Many of Wedgwood's most popular products were produced in large numbers in plain shapes, which were then embellished by designers to reflect current trends. Other more specialist products were produced in short, highly varied batches, quickly changing colour, fashion, style, and price as the market dictated. He subcontracted the manufacture of some products and their engraving to reduce his own inventory. When orders exceeded his production capacity, he outsourced from other potters. Wedgwood's innovative production system aimed to minimize proprietary risk and reduce fixed costs. He was highly aware of costs, having at one time complained that his sales were at an all-time high, yet profits were minimal. He studied cost structures and came to value economies of scale, trying to avoid producing one-off vases 'at least till we are got into a more methodicall way of making the same sorts over again'.

Wedgwood was an innovator in the way work was organized. His organizational innovations were introduced into an essentially peasant industry, with primitive work practices. When Wedgwood founded his main Staffordshire factory, Etruria, he applied the principles of the division of labour espoused by his contemporary, Adam Smith. Replacing previous craft production techniques, where one worker produced entire products, specialists concentrated on one specific element of the production process to enhance efficiency. Craftsmanship improved, allowing artists, for example, to improve the quality of designs, and innovation flourished. One of his proudest boasts was that he had 'made artists of mere men'.

Wedgwood paid slightly higher wages than the local average and invested extensively in training and skills development. In return, he demanded punctuality, introducing a bell to summon workers and a primitive clocking-in system, fixed hours, and constant attendance; high standards of care and cleanliness; avoidance of waste; and a ban on drinking. Wedgwood was conscious about health and safety, especially in relation to the ever-present dangers of lead poisoning. He insisted on proper cleaning methods, work attire, and washing facilities.

As a business innovator, Wedgwood created value by engaging with external parties in a number of ways. He innovated in sources of supply and distribution, astutely used personal and business partnerships to advantage, and introduced a remarkable number of marketing and retailing innovations.

Wedgwood sought the best-quality raw materials from wherever he could find them. In what today would be called 'global sourcing', he purchased clay from America in a deal struck with the Cherokee nation, from China, and the new colony in Australia.

He had a wide range of friends with very diverse interests upon whom he drew in his business dealings. Wedgwood belonged to a group of similarly minded polymaths who became known as the Lunar Men because of their meeting during the full moon. Along with Wedgwood, this group comprised a core of Erasmus Darwin, Matthew Boulton, James Watt, and Joseph Priestley. The friendship and business partnership with Boulton was particularly influential on Wedgwood's thinking about work organization, as he observed the efficiency, productivity, and profitability of the Boulton and Watt factory making steam engines in Birmingham. Jenny Uglow's book on the Lunar Men argues that they were at the leading edge of almost every movement of their time, in science, in industry, and in the arts. She evocatively suggests that: 'In the time of the Lunar men, science and art were not separated, you could be an inventor and designer, an experimenter and a poet, a dreamer and an entrepreneur all at once.'

Although Wedgwood had somewhat contradictory views on the ownership of intellectual property, he encouraged collaborative research and was a proponent of what today would be called 'open innovation'. In 1775, he proposed a cooperative programme with fellow Staffordshire potters to solve a common technical problem. It was a plan for what was the world's first collaborative industrial research project. The scheme failed to get off the ground, but it illustrates a desire to use a form of organization that was not again explored for over a century.

Wedgwood was the first in his industry to mark his name on his wares, denoting ownership of the design, but he disliked patents, and only ever owned one. Speaking of himself, he explains his approach:

> When Mr. Wedgwood discovered the art of making Queen's ware...he did not ask for a patent for this important discovery. A patent would greatly have limited its public utility. Instead of one hundred manufactures of Queen's ware there would have been one; and instead of an exportation to all quarters of the world, a few pretty things would have been made for the amusement of the people of fashion in England.

The period of the Industrial Revolution was one of great optimism as well as social upheaval. Consumption and lifestyle patterns changed as industrial wages were paid and new businesses created novel sources of wealth. The population of England doubled from around five million in 1700 to ten million in 1800. Until the 18th century, English pottery had been functional: mainly crude vessels for storing and carrying. Pots were crudely made, ornamented in an elementary way, and glazed imperfectly. The size and sophistication of the market developed throughout the 18th century. Stylish table accessories were in huge demand in the burgeoning industrial cities and increasingly wealthy colonies. Drinking tea, as well as the more fashionable coffee and hot chocolate, joined the traditional British pastime of imbibing beer as a national characteristic.

Wedgwood sought to meet and shape this burgeoning demand in a number of ways. Initially he sold his completed wares to merchants for resale, but he opened a warehouse in London, followed by a showroom that took direct orders. Browsing customers commented on the wares on display, and Wedgwood took particular note of criticisms of uneven quality, explaining his devotion to researching how to achieve better consistency. The showroom, run by Wedgwood's close friend, Thomas Bentley, became a place for the fashionable to be seen, and major new collections were visited by royalty and aristocracy. Bentley expertly interpreted new trends and tastes, informing design and production plans back in Staffordshire.

Wedgwood assiduously sought patronage from politicians and aristocrats: what he called his 'lines, channels, and connections'. He produced a 952-piece dinner service for Catherine the Great, Empress of Russia, shamelessly using her patronage in his advertising. His view was that if the great and the good bought his products, the new middle classes, merchants, and professionals, and even some of the wealthier lower classes, artisans and tradespeople, would aspire to emulate them.

An astonishing number of retailing innovations were introduced by Wedgwood and Bentley, including the display of wares set out in full dinner service, self-service, catalogues, pattern books, free carriage of goods, money-back guarantees, travelling sales people, and regular sales, all aiming 'to amuse, and divert, and please, and astonish, nay, and even to ravish the Ladies'. Jane Austen wrote of the pleasure of the safe delivery of a Wedgwood order.

Wedgwood's international marketing efforts were pioneering. When he started his business, it was rare for Staffordshire pottery to reach London, let alone overseas. To sell in international markets, he again used the strategy of courting royalty by using his English aristocratic connections as ambassadors. By the mid-1780s, 80 per cent of his total production was exported.

Products were not sold on the basis of low prices. Wedgwood's products could be two or three times as expensive as his competitors'. As he put it, 'it has always been my aim to improve the quality of the articles of my manufacture, rather than to lower their prices'. He was contemptuous of price cutting in the pottery industry, writing to Bentley in 1771:

> the General Trade seems to me to be going to ruin on the gallop...low prices must beget a low quality in their manufacture, which will beget contempt, which will beget neglect, and disuse, and there is an end of the trade.

Wedgwood's innovations extended into many other areas. He expended substantial efforts in building the infrastructure supporting the manufacture and distribution of his products and those in his industry. He devoted significant amounts of time and money to improving communications and transportation, especially with the ports that supplied raw materials and provided his routes to market. He promoted the development of turnpike roads and became centrally engaged in the construction of major canals. He actively lobbied the government on trade and industry policy and helped form the first British Chamber of Manufacturers.

Wedgwood's legacy extended well beyond his company. He had an enormous impact on the Staffordshire Potteries more generally, in what today might be called an innovative 'industrial cluster'. Pottery production in Staffordshire developed rapidly due to the efforts of numbers of firms, such as Spode and Turner, but Wedgwood was the acknowledged leader of the industry.

His 19th-century biographer, Samuel Smiles, wrote of the change from the 'poor and mean villages' brought about by Wedgwood's innovations:

> From a half-savage, thinly peopled district of some 7000 persons in 1760, partially employed and ill remunerated, we find them increased, in the course of some twenty-five years, to about treble the population, abundantly employed, prosperous, and comfortable.

Wedgwood's contributions to public life included improving the education, health, diet, and housing of his employees. Etruria's seventy-six homes were, in their time, considered a model village.

Wedgwood built a dynasty. He inherited £20 from his father, and when he died he left one of the finest industrial concerns in England with a personal worth of £500,000 (around £50 million at present prices). Wedgwood's children used their good fortune well. One son created the Royal Horticultural Society and another contributed centrally to the development of photography. Wedgwood's wealth was used to great effect to fund the studies of his grandson, Charles Darwin.

The Wedgwood case raises a number of core issues that we shall be examining in this *Very Short Introduction* and reveals the approach to innovation we shall be taking. We focus on the organization, the mechanism for creating and delivering innovation. The individuals and their personal connections whose importance is so clearly shown in the Wedgwood case will be discussed here only to the extent to which they contribute to organizational outcomes. We do not discuss the meanings of innovation for us individually. Nor do we adopt the perspective of the user of innovation, although we shall argue that innovative organizations need to try to understand how innovations are consumed and for what purpose. With this observation in mind, Wedgwood shows us that innovation occurs in many forms and ways. It occurs in what organizations produce: their products and services. It is found in the ways in which organizations produce: in their production processes and systems, work structures and practices, supply arrangements, collaboration with partners, and very importantly how they engage with and reach customers. Innovation also happens in the context within which organizations operate: for example, in regional networks, supportive infrastructure, and government policies.

Wedgwood illustrates an enduring truth about innovation: it involves new combinations of ideas, knowledge, skills, and resources. He was a master at combining the dramatic scientific, technological, and artistic advances of his age with rapidly changing consumer demand. Gladstone said: 'He was the greatest man who ever, in any age or in any country, applied himself to the important work of mixing art with industry.' The way Wedgwood merged technological and market opportunities, art and manufacturing, creativity and commerce, is, perhaps, his most profound lesson for us.

Chapter 2
Joseph Schumpeter's gales of creative destruction

All economic and social progress ultimately depends on new ideas that contest the introspection and inertia of the status quo with possibilities for change and improvement. Innovation is what happens when new thinking is successfully introduced in and valued by organizations. It is the arena where the creation and application of new ideas are formally organized and managed. Innovation involves deliberate preparations, objectives, and planned benefits for new ideas that have to be realized and implemented in practice. It is the theatre where the excitement of experimentation and learning meets the everyday realities of limited budgets, established routines, disputed priorities, and constrained imagination.

There are a great many ways of understanding innovation that provide a wide range of rich insights and perspectives. The variety of different analytical lenses used depends on the particular innovation issues being studied. Some analyse the extent and nature of innovation: whether any change is incremental or radical, how it sustains or disrupts existing ways of doing things, and if it occurs in whole systems or their components. Other analyses are concerned with how the focus of innovation changes over time, that is, from the development of new products to their manufacture, their patterns of diffusion, how particular design configurations, such as mobile phones and online banking services, come to dominate in the market, and how to appropriate value from innovation.

Defining innovation

The relatively simple definition of innovation we use—ideas, successfully applied—helps differentiate it from invention and creativity, which can valuably contribute ideas prior to their application. Yet it still has very broad meanings, which can be helpful in as much as it can usefully cover a wide range of activities and is confusing for the same reason—the word can be used promiscuously. Even our straightforward definition raises questions. What is 'success'? Time is influential, and innovations may be initially successful and eventually fail, or vice versa. What does 'applied' imply? Is it applied within a single part of an organization, or diffused internationally among a massive group of users? What and who are the sources of 'ideas'? Can anyone lay claim to them, especially as they inevitably combine new and existing thinking?

Typologies of innovation also face difficulties because of blurred boundaries and overlaps between categories. Innovation occurs in products, for example in new cars or pharmaceuticals; and services, for example in new insurance policies or in health monitoring. But many service firms describe their offerings as products, such as new financial products. Innovation occurs in operational processes, in the way new products and services are delivered. These processes may take the form of equipment and machinery, which are the providers' products, and logistics in the form of transportation, which are providers' services.

There are some similar definitional problems when thinking about levels of innovation. A minor innovation for one organization may be substantial for another. It is difficult in practice to draw boundaries between different categories, because organizations judge their innovativeness within their own specific circumstances.

Most innovations are incremental improvements—ideas used in new models of existing products and services, or adjustments to organizational processes. They might include the latest versions of particular software packages, or decisions to add more representatives from the marketing department to development teams. Radical innovations change the nature of products, services, and processes. Examples include the development of synthetic materials, such as the Gore-Tex fabric that is rain- and wind-proof yet breathable, and decisions to use open-source software to encourage community development of new services, rather than doing it proprietarily. At the highest level, there are rarer, periodic transformational innovations, which are revolutionary in their impact and affect the whole economy. Examples would be the development of oil or photovoltaics as energy sources, or the computer or Internet.

We think about innovation as ideas, successfully applied in organizational outcomes and processes. Innovation can be thought of as practical and functional: the outcomes of innovation are new products and services, or they are the processes supporting innovation that occur in departments such as research and development (R&D), engineering, design, and marketing. Innovation can also be thought of more conceptually: the outcomes of innovation are enhanced knowledge and judgement, or they are the processes that support the capacity of organizations to learn. Innovation can be conceived as a way of providing options when facing an uncertain future.

We have chosen to focus on innovations other than those described as 'continuous improvement' that tend to be routine and highly incremental in nature. Although these small improvements are cumulatively important, our concern lies rather with ideas that stretch and challenge organizations as they attempt to survive and thrive. By concentrating on innovations beyond the ordinary that occur in both the outcomes of efforts and the processes that produce them, we capture a great degree of what is generally understood to be innovation.

Importance of innovation

The reason why innovation is so important has to be seen in the context of the relentless demands made of contemporary organizations as they face the challenges of a complex and turbulent world. Innovation is crucial for their continuing existence as they struggle to adapt and evolve to deal with constantly changing markets and technologies.

In the private sector, the threat of new competitors in globalized markets is ever present. In the public sector, the demand for efficiencies and enhanced performance is continual, as governments attempt to manage calls for expenditure to improve the quality of life, and address emerging problems such as cybersecurity or extreme weather events, that exceed their incomes. Charities and non-governmental agencies need constantly to think of novel and more effective ways of attending to the problems they address. The motivation to innovate in all organizations is stimulated by the knowledge that if they are not capable of innovation, others are: new players that may threaten their very existence. Simply, if organizations are to progress—to develop and grow, become more profitable, efficient, and sustainable—they need successfully to implement new ideas. They have to be continually innovative. As the economist Joseph Schumpeter (Figure 2) put it, at its most blunt, innovation 'offers the carrot of spectacular reward or the stick of destitution'.

One of the features of innovation is that it can be found in every organization. Although the cost of innovation can be very high—it can according to some estimates cost between $1.5 and $2.5 billion to develop a new pharmaceutical—new ideas can be successfully implemented cheaply. It is not only high-tech firms

2. Joseph Schumpeter placed innovation centrally in his theory of economic development.

making semiconductors or working with biotechnology that rely on innovation in their businesses, it is all parts of the economy. Insurance firms and banks continually search for new ideas for services for customers; shops use sophisticated digital infrastructure to manage their ordering and stock; farms use new seeds, fertilizers, and irrigation technologies, satellites can assist the optimization of their planting and harvesting, and new uses are being made of their products, such as biofuels and health-promoting functional foods. Innovation is found in construction, in new materials and building techniques; in packaging that keeps food fresher; and in clothing firms introducing new designs more quickly and cheaply. Innovation is sought by public services, in health, transportation, and education. Charities can increase their funds through crowdsourcing campaigns. While one might not wish for too much innovation in some areas, such as among the firms investing our pension funds, generally the business or organization that doesn't benefit from using new ideas is rare indeed.

Challenges

The challenges of innovation are immense. Many people are uncomfortable with the changes brought about by innovation. Especially when it is broad-ranging, innovation can have negative effects on employees, inducing uncertainty, fear, and frustration. Organizations have social contracts by which their members develop loyalty, commitment, and trust. Innovation can disrupt this contract by redistributing resources, altering the relationships between groups, and emphasizing the ascendancy of one part of the organization to the disadvantage of others. It can disturb the technical and professional skills people acquire over many years, and with which they strongly identify themselves. This means it is inseparable from the exertion of power and resistance to it.

Most attempts at innovation fail. History is littered with unsuccessful attempts to apply the—often very good—new ideas of individuals and organizations. The ill-fated development of a cost-effective battery electric car with significant environmental benefits in the USA in the 1990s is illustrative of the way innovation can provide a serious threat to established interests. A coalition of political and business interests combined to prevent this new idea reaching the market at that time. Although the product was popular with consumers, it had to compete with the concerns of the established energy infrastructure, oil companies, and petrol distribution networks, and massive existing automotive industry investments in petrol engine car manufacture and maintenance.

Organizations simultaneously need to do things that allow them to operate in the short term, exploiting their existing know-how and skills, and explore new things that will develop capacities to support their continued long-term existence in a changing world. These demand different and sometimes conflicting behaviours and practices. Indeed, organizations are occasionally confronted with the paradox of needing to apply new ideas threatening to the practices that have created their current successes. If generals are said to fight the last rather than the current war, managers rely on ways of doing things that contributed to their organizations', and their own, past progress, rather than ways that will deal more effectively with the future. Since Edison established the first company dedicated to producing innovations at the turn of the 19th century, many different ways of structuring the creation and use of ideas have periodically been favoured. As the business environment has changed, the large, centralized corporate R&D laboratory is no longer used as often as in the past. The search for ways of balancing routines with innovation is constant.

Organizations rarely innovate alone: they do so in association with others, including their suppliers, customers, and communities of users. They innovate in particular regional and national contexts. Access to innovation-supporting skills and university research, for example, often has a local dimension, as seen in the case of Silicon Valley and other international centres of innovation. Government policies and regulations affect innovation, as do national financial and legal systems that influence issues such as the availability of risk-taking investment capital, the creation of technical standards, and protection of intellectual property rights. Availability and cost of infrastructure for communications and transportation matter greatly. These factors add to the complexity, and hence unpredictability, of innovation, as innovators are never completely masters or mistresses of their destiny. They also point to the essentially idiosyncratic nature of innovation: each innovation occurs in its own particular set of circumstances.

In all the major elements of contemporary economies—in the services, manufacturing, and resources industries, and in the public and third sector—progress depends upon owning or accessing and using knowledge and information. Being competitive and efficient relies on being innovative with all the resources organizations possess: their people, capital, and technology, and the ways they connect with those contributing to and using what they do.

As well as the challenges in developing and implementing innovation for people and organizations it is also necessary to consider the broader social and political challenges associated with innovation. Employment levels and the nature of jobs are profoundly affected by innovation. It has given us weapons of mass destruction and caused immense environmental damage. For all the benefits of the Internet, it has also aided terrorism, child exploitation, and online bullying. The broader social impacts of innovation are a subject to which we shall return.

Innovation thinking

The American economist William Baumol argued that virtually all of the economic growth that has occurred since the 18th century is ultimately attributable to innovation. The successful application of ideas has been recognized within industry as the primary source of its development since this time.

The 18th century also saw the beginning of the study and recognition of the importance of the relationships between organization, technology, and productivity with the publication of Adam Smith's *Wealth of Nations* in 1767. Smith produced his now famous analysis of the importance of the division of labour in a pin factory which was so influential on Wedgwood's factory. Smith showed how specialization in specific manufacturing processes in pin production significantly increased the productivity of the workforce compared to when individuals produced each pin themselves.

A man alone, even with the 'utmost industry', could produce 1 to a maximum of 20 pins a day, yet with a division of labour, 'very poor' labour 'indifferently accommodated with the necessary machinery', could produce 4,800 'when they exerted themselves'.

A century later, Karl Marx was highly aware of the significance of innovation, but more concerned with its negative consequences. In the first volume of *Capital*, he declared:

Modern industry never views or treats the existing form of a production process as the definitive one ... By means of machinery, chemical processes and other methods, it is continually transforming not only the technical basis of production, but also the functions of the worker and the social combinations of the labour process.

The possibilities of technological change, Marx argued, were contradicted by its use under capitalism, which inevitably led to the suppression of workers. Capitalism, he contended, subordinated workers to machines, but he believed technology held the possibility of liberating them from the burden of mechanical and repetitive work and enriching social relations.

Marx's emphasis on the strong social dimensions to technological development and use is a recurrent theme in research into the history of innovation. Study of the development of automated machine tools in the USA, for example, illustrates how often technology is shaped by dominant social forces. The automated, or numerical, control of machine tools, such as lathes, could have been configured in various ways to give the machine's operator more or less discretion over how it was used. The technology was constructed in such a way that control resided in engineering planning offices, not with their operators. This was less economically efficient, but complied with the expectations of the major customer for the new technology, the US Air Force, and hence reflected existing power structures.

At a more aggregate level, all the past revolutions in technology—in steam power, electricity, automobiles, information and communications technology—have required enormous adjustment and adaptation in industry and society. The economists Christopher Freeman and Carlota Perez show how in history the diffusion of new technologies since the Industrial Revolution required massive structural adjustments in industry and society, and also in the legal and financial framework, education and training systems for new skills and professions, new management systems, and new national and international technical standards.

The importance of clever 'human capital' has long been recognized. Observing the development of German industry in the mid-19th century, the political scientist Friedrich List declared that national wealth is created by intellectual capital: the power of people with ideas. In 1890, the British economist Alfred Marshall noted that knowledge is the most powerful engine of production available to economies. An economic theorist who unusually kept his feet on the ground by regularly visiting companies, Marshall celebrated the importance of innovation and is especially remembered for his analysis of the benefits of the 'clustering' of progressive firms in 'industrial districts'.

If any economist lays claim to be the first to include innovation centrally within their theory of development, it is Joseph Schumpeter (1883–1950), who remains today one of the most influential thinkers on the subject. A complex man with a rich history, including being one-time finance minister in Austria, director of a failed bank, and Harvard professor, Schumpeter argued that innovation unleashed the 'gales of creative destruction'. It arrives in a great storm of revolutionary technologies, such as oil and steel, that fundamentally change and develop the economy. Innovation is creative and beneficial, bringing new industries, wealth, and employment, and at the same time is destructive of some established firms, many products and jobs, and the dreams of failed entrepreneurs. For Schumpeter, innovation is essential for competitive survival:

> the competition from the new commodity, the new technology, the new source of supply, the new type of organization...competition which commands a decisive cost or quality advantage and which strikes not at the margins of the profits and the outputs of the existing firms but at their foundations and their very lives...

Schumpeter's views on the primary sources of innovation altered during his lifetime, reflecting changes in the practices in industry. His early 'Mark I' model, published in 1912, celebrated the importance of individual, heroic, risk-taking entrepreneurs. His 'Mark II' model, by contrast, published thirty years later, advanced the role of the formal, organized innovative efforts in large companies. It was during this period that the modern research laboratory became firmly established, initially in the chemical and electrical industries in Germany and the USA. By 1921, there were more than 500 industrial research laboratories in the USA.

Five models. One of the first and most influential studies of the relationship between scientific progress and industrial innovation was conducted immediately after the Second World War by Vannevar Bush, the USA's first presidential science advisor. In his report *Science: The Endless Frontier*, Bush advocated a national policy for open-ended research on a massive scale. The book proved popular; it was serialized in *Fortune* magazine, and Bush appeared on the front page of *Time*. The view that investments in research held the solutions to most seemingly intractable problems was a legacy of Bush's association with the Manhattan Project to develop the atom bomb, which, to the minds of many, successfully curtailed the war in the Pacific. Although it took a simplistic interpretation of Bush's report, the view that all product and process innovations are founded in painstaking basic research became the fundamental precept of the *science push* model of innovation, a perspective that remains popular with many in the scientific research community to this day.

An alternative view, which emphasized the importance of market demand as the primary source of innovation, emerged in the 1950s and 1960s. This resulted from a number of factors, including studies that showed that in sectors such as the military, technological outcomes resulted more from the demands of its users than from any scientifically predetermined configurations. At the same time, there was a growth of large, corporate planning offices with belief in the conceit that sufficient market research could identify what was required of new science and technology to meet consumer needs. This mirrored the rise of social science at the time with its claims to predictive powers. Counter to the post-war enthusiastic embrace of science and technology, social movements—such as Ralph Nader's car safety campaign in the 1960s, developed in response to dangerous car designs—began to question the use to which they were being put and demand greater attention to consumer needs. In housing, research into the demographics of the baby-boomer generation led to 'predict and provide' strategies internationally, where innovation was sought to help satisfy growing demand. This view became known as the *demand-pull* model of innovation.

Both these models of innovation are linear in their progression: research leads to new products and processes introduced into the market, or market demand for new products and processes leads to research to develop them. But increasing volumes of research conducted in the 1970s questioned the assumption of linearity. Pioneering studies, such as Project SAPPHO at the University of Sussex, UK, found differences between sectors: for example, the chemical industry innovated differently from the scientific instruments industry. And, furthermore, the pattern of innovation changed over time. Abernathy and Utterback at MIT developed the theory of product life cycles, with high levels of innovation in the development of products occurring initially, then reducing in scale and being replaced by high levels of innovation focusing on their application and their processes of production. Innovation was seen not to be unidirectional, but more iterative, with feedback loops.

The organizational and skills issues underlying this *coupling* model of innovation came to the fore in the 1980s, driven primarily by the remarkable success of Japanese industry during this period. A study of the car industry at the time showed Japanese auto manufacturers were twice as efficient as their international competitors in every measure of innovative performance, such as how long it takes to design and make a car. The explanation for this was an approach described as 'lean production', which contrasted with the mass-production techniques used in other countries. Mass production, typified by Henry Ford, is based on assembly lines producing standardized products. 'You can have any colour Model T car you want, as long as it is black', as Ford is reputed to have said. Lean production introduced greater flexibility into the assembly line, allowing a broader range of products to be made. It includes a system of relationships with suppliers of components that allow them to deliver 'just-in-time' to be assembled, thereby reducing the cost of holding inventory, and increasing the speed of response to market changes. Lean production also entailed an obsessive concern for quality control, which in many areas became the responsibility of shop-floor workers.

When analysing the differences between the way Japanese and Western firms organized themselves to innovate, the metaphors were used at the time of the former playing rugby (although netball is similarly suitable) and the latter running a relay race. In the West, innovation entailed one part of the company, say R&D, beginning the process, running with it for a while, then handing it over to another, say engineering, which similarly worked on it before passing it over to manufacturing and then on to marketing.

This linear process was considered hugely wasteful to Japanese firms, with the likelihood of significant miscommunication as projects moved from one part of the company to another.

A rugby or netball player metaphor can be used, as their games involve the simultaneous combination of different kinds of players, with various skills and abilities, some big and tall, but generally slow, and some smaller, skilful, and fast, all working to the same objective. All parts of the organization were combined in innovation activities.

Collaboration between, as well as within, innovative Japanese companies was a feature of their 1980s' success story. As well as the extensive collaboration between customers and suppliers in the same industrial groups—Keiretsu—the Japanese government also encouraged collaboration between competing firms. The Fifth Generation Computer program, for example, attempted to encourage the usually highly competitive computer manufacturers to cooperate around shared research agendas. This *collaborative* model of innovation strategies and public innovation policies was also enthusiastically pursued in Europe in information technology (IT) and the USA in semiconductors.

By the 1990s, Roy Rothwell, one of the founders of innovation research, began to identify a number of changes occurring in the strategies firms were using to innovate and the technologies they were using to support it. He argued firms were developing innovation strategies that were highly integrated with their partners, including 'lead customers', demanding users, and co-developers of innovation. Also important, he contended, was the use of new digital technologies, like computer-aided design and manufacturing, that brought different parts of the firm together when developing innovations, and helped link external parties into internal development efforts. Rothwell called this the *strategic integration and networking* model of innovation. The trend towards greater strategic and technological integration in support of innovation is one that continues with the use of massive computing power, the Internet, and new visualization and digital technologies such as artificial intelligence (AI).

These models of the innovation process have their antecedents in an industrialized economy where innovation predominantly occurs in the manufacturing industry. We are now in economies where services dominate, accounting for around 80 per cent of gross domestic product in most developed nations. Much value-added in economies lies with merger of services and manufacturing, and the blending of digital and physical systems. Economies based on tangible, physical objects that can be measured and seen have been transformed into those where outputs are weightless and invisible. Furthermore, as the global financial crisis that emerged in 2008 shows, we live in an era of extraordinary turbulence and uncertainty where any established formulae and prescriptions are likely to be tested by emerging and unforeseen circumstances. The models of innovation in the future will be far more organic and evolutionary where the sources of innovation are unclear, the organizations involved initially unknown, and the outcomes highly constrained by unpredictability. In these circumstances, it will be valuable to assess whether or not anything we know of the past might be a guide to the future. It will also be useful to understand how theory of innovation might help.

Theory

There is no single, unified theory of innovation. There are partial explanations from, for example, economics, political science, sociology, geography, organizational studies, psychology, business strategy, and from within 'innovation studies', which draws on all these disciplines. This is to be expected given innovation's multiple influences, pathways, and outcomes. The utility of various theories will depend on the particular issues being explored. Theories from psychology may be more useful when the subject is an individual innovator, business strategy when it is organizational innovation, and economics when it is national innovation performance. It is important to consider theories of innovation not only to explain contemporary issues, significant as they are, but also to enlighten its future use in helping to deal with major social, economic, and environmental concerns.

Over recent years, there has been an emergence of a number of perspectives that share common ground in their theories of innovation. These include evolutionary economics at the macro level and 'dynamic capabilities' frameworks for business strategy at the micro level.

The challenge for any theory of innovation is that it has to explain an empirical phenomenon that takes many guises. It also has to encompass its complexity, dynamism, and uncertainty, often compounded by the way innovation results from the contribution of many parties with occasionally divergent and not fully established agendas. In this way, innovation has emergent properties: it results from a collective process whose outcomes may not be known or expected when it begins.

Evolutionary economics—with a Schumpeterian legacy—sees capitalism as a system that produces continuous variety in the new ideas, firms, and technologies created by entrepreneurs and the innovative activities of research groups. Decisions by companies, consumers, and governments make selections from within this variety. Some selections are successfully propagated and fully developed into new organizations, businesses, and technologies that provide the basis and resources for future investments in creating variety. Much of the variety and selections that occur are disruptive or fail to be propagated, so the evolutionary development of the economy is typified by significant uncertainty and failure.

Dynamic capabilities theory includes the ways firms search for, select, configure, deploy, and learn about innovations. Its focus is on the skills, processes, and structures that create, use, and protect intangible and difficult to replicate assets, such as knowledge. This approach to strategy reflects the continual dynamism of technology, markets, and organizations where the capacity to sense threats and realize opportunities—when information is constrained and circumstances unpredictable—is the key to sustainable corporate advantage.

These theoretical explanations for innovation embrace complexity and emergent circumstances. They incorporate the messy realities of innovation found in economies where there is constant change and adaptation, and when the strategies of firms are often experimental.

Time

There has to be a time dimension to any understanding of innovation. Whether considering outcomes, what happened, or processes of innovation, how it came about, it is necessary to know the period in which they occurred. Comparisons to what existed before the innovation determine the extent of novelty.

If an innovation is ahead of its time, as perhaps might be argued to have happened in the case of the battery electric car in the 1990s, no matter how much effort is expended, it will not gain the momentum needed for its wide diffusion and sustained growth. If an innovation takes too long to be developed, it may fail because a superior or cheaper idea emerges. Sometimes markets and technologies shift quickly, and rapidly move on from what at one point seemed like a good idea. Innovative organizations therefore have to think about timescales of new ideas. They can do this by considering their position based on past innovation diffusion patterns, and use tools and techniques to speed up innovation through formal project management techniques that progressively decide on the levels of resources needed. Returns on innovation investment are planned over periods of years, and decisions are made to invest if they pay back suitably over an acceptable time period. Risk is managed by attempts to reduce how long it takes to develop and introduce innovation. Speed is usually, but not always, seen to be a benefit. Compressing time reduces the chance of being caught up by competitors or wastefully squandering resources. Moving too quickly, however, leads to mistakes and failures to learn from them.

Short-term horizons are appropriate for incremental innovation, but long-term perspectives are needed to provide a wider view on where, why, and how radical innovation has occurred or failed. Understanding the relationships between scientific discovery, innovation, and societal changes requires deep historical interpretation.

Innovators—as we shall see in the case of Edison in Chapter 5—can improve their future chances of success by creating options that allow different potential paths to be followed, delaying decisions that do not need to be made until a later date when their consequences may be clearer. By organizing and equipping themselves for unforeseen eventualities, innovators can change course, or recalibrate timetables. As Louis Pasteur observed about scientific discovery through experimentation, 'chance favours the prepared mind'.

Rates of innovation and diffusion vary considerably between different business sectors. In pharmaceuticals, for example, it usually takes between twelve and fifteen years to bring a new drug to market, but new digital services can grow large within months. It took Facebook ten years to move from a university dormitory to become more valuable than the Bank of America. Organizations can make strategic choices on whether they should attempt to lead innovation in their sector or follow others. Sometimes leaders have the best opportunity for reaping the greatest rewards from their ideas. The chemical company DuPont, for example, has consistently led other firms in bringing new products to market for over a century. But 'first-mover advantage' may be difficult to capture and sustain. It often brings greater risks, as the market may not be fully formulated, and higher costs may be accrued to stimulate demand.

Other organizations choose to learn from leaders, emulating innovations that appear to work well, and avoiding any pitfalls they have observed. Fast-followers can receive huge rewards, as Microsoft has done for consistently reacting quickly to the innovations of others who have borne initial risks. Many organizations do not have the skills or resources to be first-movers or fast-followers. They may, nevertheless, benefit from innovation that improves upon, adapts, or extends products, processes, or services. Whatever its position as an innovator, and whichever strategy is pursued, its ability to appreciate the time dimension is likely to have a significant bearing on its performance.

Understanding how innovations are consumed and diffused

The impact of innovations depends on the extent to which they are diffused. The classic study of this subject, *Diffusion of Innovations*, was written by Everett Rogers in 1962, who states on the first page of the book: 'getting a new idea adopted, even though it has obvious advantages, is often very difficult. Many innovations require a lengthy period, often of many years, from the time they become available to the time they are widely adopted.' Decisions on whether or not to take up an innovation are not instantaneous, but a process that occurs over time, consisting of a series of different actions. A range of prior conditions bring consumers into the process in the first place, including their previous experiences, existing needs and problems, norms of their social systems (e.g. their social groups), and general 'innovativeness'. Innovation adoption is therefore strongly bounded by the social context in which it occurs. For Rogers, the innovation-decision is a social and psychological process as much as an economic one.

The implication for organizations developing innovative products and services is that they need to understand how their offerings are consumed and what meaning is attached to them. The Sony Walkman—a mobile music playing device using tape cassettes—was originally created for urban youths listening to their music. It had two headphone jack sockets for listening with friends simultaneously, because solitarily listening to music in public places was believed to be impolite: the assumption being that the pleasure of music was something naturally shared. However, many more people bought a Walkman, young and old, particularly those engaging in outdoor activities such as jogging and bike riding; and people listened to the music individually. The way people used the Walkman was more personal than shared. The Sony Walkman II had a new design with only one headphone jack socket. The image of outdoor activities was also incorporated into the product's advertising. The Sony Walkman II was a great success because the ways it was consumed were better understood by designers.

Japanese sociologist, Ritsuko Ozaki, studied the reasons why consumers bought hybrid cars, specifically the Toyota Prius. She found their consumption is not only about reducing petrol usage but also about the self-expression of being part of a green community. Financial concerns—initial and subsequent running costs—were found to be centrally important, with issues of fuel economy and reduced road taxes being especially valued by consumers. Factors such as size, comfort, quietness, and ease of use added to the practical, rational, and utilitarian dimensions of purchase decisions. The reputation of the company for reliability and consumers' past experience of driving Toyota cars or other hybrid cars are influential. Also rated highly by purchasers are the car's perceived environmental benefits and compatibility with their environmental values/beliefs. These reflect personal and social expressions through consumption. Personal interest in technology is also highly relevant. Some people are intrinsically attracted to technology and have a positive attitude towards technical novelty, such as the combination of electricity and petrol engine in hybrid cars.

The consumption of innovation, therefore, moves beyond financial, practical, and aesthetic reasons, and includes social pressures and norms and personal attitudes towards novelty and new experiences.

When taking a wider societal perspective on the diffusion of innovations, it has long been argued that radical technological innovations require extensive adjustments in the society and economy of which they are a part. These can range from new industrial structures to the need for novel skills, industrial relations, and regulations, all of which can take many years if not decades. Yet one of the most extraordinary business phenomena over recent years has been the rapid growth of social media platforms such as Twitter and Instagram, ecommerce companies such as Alibaba and eBay, and 'shared economy' organizations such as Uber and Airbnb. What this implies is that whereas many complex innovations, such as the microgeneration of electricity in the home, will take time for the various aspects of its technology, economics, and social acceptance to co-evolve, when a large latent social demand is met—i.e. for cheaper accommodation and transportation—diffusion of radical ideas can happen quickly.

Chapter 3
London's wobbly bridge: learning from failure

Schumpeter's analysis of innovation being a process of creative destruction implies the outcomes of innovation can simultaneously be positive and negative. It both creates and destroys wealth and jobs. Innovation profoundly affects us all by creating new industries, firms, and products, as seen in the new industry established by Wedgwood. It is seen in services such as discount airlines, and infrastructure such as airports. It improves productivity and quality of life in the form, for example, of new pharmaceuticals, means of transportation, communication, entertainment, and greater variety in and accessibility of food. It has helped lift millions out of poverty, especially over recent decades in Asia. Jobs can be more creative, interesting, and challenging as a result of innovation. But the successful application of ideas can also have profoundly adverse consequences. Nations and regions get left behind when they are not as innovative as their competitors, and increasing disparities in wealth result. Jobs can be deskilled, job satisfaction decreased, and unemployment increased, because of innovation. Innovation has given us the environmental consequences of the internal combustion engine and chlorofluorocarbons, cluster bombs, and chemical weapons, and the toxic results from the complex financial instruments behind 2008's global financial crisis.

Predicting the adverse consequences of innovation can be as challenging as foreseeing its positive effects: these are unpredictable and can be mixed. On the positive side, the internal combustion engine democratized travel, chlorofluorocarbons in refrigerators improved nutrition, financial innovations gave us the security of better life insurance and pensions. But the occasionally ambiguous nature of the results of innovation is seen in the case of failure. Most attempts at innovation fail, and there is a highly skewed distribution in its returns, but failure is itself an important outcome, and it is to this that we now turn.

Failure

Innovation is risky, as, for example, innovators have to consider:

- Demand risk—how big will the market be for a new product or service? Will new competitors emerge?
- Business risk—is appropriate finance available to meet the costs of innovation? What effect will an innovation have on organizational reputation and brands?
- Technology risk—will a technology work, is it safe, and how does it complement other technologies? Will better competing technologies emerge?
- Organization risk—are the right management and organizational structures being used? Are the necessary skills and teams available?
- Network risk—are the right collaborative partners and supply chains in place? Are there important gaps?
- Contextual risks—how volatile are government policies, regulations, and taxation, and finance markets?

In theory, risk can be measured and managed by making assumptions on probabilities, although there are always dangers in assuming the past can predict the future. Uncertainty, on the other hand, has a truly unknown outcome and cannot be measured, and its management depends on decisions based on deep experience and intuition. It is because of risks and uncertainties that there is so much failure in innovation, but at the same time, they provide an incentive. If there were no risk and uncertainty, and therefore everyone could innovate easily, then innovation would provide little advantage over competitors.

Failures also provide valuable opportunities for future improvements, as seen in the highly embarrassing case of London's Millennium Bridge. Linking the Tate Gallery and St Paul's Cathedral, this was the first footbridge to be built across the River Thames for more than a hundred years. It is an extraordinary engineering, architectural, and sculptural achievement—a design of such beauty it has been described as a 'blade of light' across the river (see Figure 3). The bridge opened on 10 June 2000, when between 80,000 and 100,000 people walked across it. When large groups of people were crossing, however, it became noticeably and increasingly unsteady, and it quickly gained notoriety as the 'wobbly bridge'. The bridge was closed after two days, causing immense discomfiture to all concerned.

3. The Millennium Bridge: a great success after a wobbly start.

After an intensive international effort, the cause was found and rectified. The problem, apparently, was the way men tend to walk with splayed feet, like ducks. When many of them walk in unison, unusual 'lateral excitation' occurs. Had it been a women-only bridge, there would not have been a problem. As a result of this debacle, new knowledge about bridge design was developed, and future projects will allow large numbers of men to waddle happily together over rivers.

The Millennium Bridge is an example of the way in which much progress in science, engineering, and innovation is built upon failure. As the chemist Humphry Davy said: 'The most important of my discoveries has been suggested to me by my failures.' And as Henry Ford put it: 'Failure is only the opportunity to begin again more intelligently.' Empirical evidence shows how returns to new ideas are highly skewed—there is what physicists and economists call a 'power law distribution'. Only a few academic papers, patents, products, and company start-ups are successes. In most cases, the majority of returns come from 10 per cent of innovative investments. In some areas it is even more skewed. At any one time there may be up to 8,000 potential new pharmaceuticals being researched around the world, but maybe only one or two will prove successful.

There is a strongly temporal element to failure: things deemed failures can become successful, such as the Millennium Bridge, and successes can over time turn into failures. Following its introduction in 1949, the de Havilland Comet aircraft was instrumental in creating the international commercial airline industry. The Comet was considered a highly successful product innovation until the period in the mid-1950s when the aircraft started falling out of the sky with alarming regularity. Aircraft engineers at the time knew little about metal fatigue, which was the cause of the crashes, but aircraft design improved as a result of the lessons learned from these failures.

Products may succeed technologically but fail in the market. The Sony Betamax was technically a better video recorder than its competitor, Matsushita's VHS system, but it lost the competitive battle to become the dominant design in the market. The supersonic jet airliner Concorde was a technological marvel in its time, but it sold only to the British and French governments, its joint manufacturer.

It is not always possible to judge what is going to be valuable in the future. Apple's Newton—an early personal digital assistant—is a notorious product failure. It cost more than a computer, and one memorable technical review said it was so big and heavy it could only be carried around by kangaroos. Its failure cost Apple's CEO his job. Yet ten years later and its operating system was found in the iPod, and several Newton-like features were incorporated in the iPhone.

Failure has a personal cost, and innovators have to develop strategies to deal with failure that entail personal recognition of its value for learning, reflection, and self-awareness. Similarly, organizations need to appreciate the value, and learn the lessons, of failure.

Learning

Innovation is manifested in new products, services, and processes. Less material, but no less real, are the options for the future it provides and the organizational and personal learning it encourages.

Organizations learn to do better the things they already do, learn to do new things, and learn about the need to learn. Organizations inevitably learn by doing familiar things; the more you do something, generally the better at it you become. But radical and disruptive innovations—those that involve significant breakthroughs and fracture past ways of doing things—pose great difficulties for organizations and the ways they learn. Established routines and ways of doing things actually restrain learning about these forms of innovation. Focus on the status quo produces returns that are positive, proximate, and predictable; focus on the novel produces returns that are uncertain, distant, and often negative. This produces a tendency to substitute exploitation of known alternatives for the exploration of unknown ones. Radical innovation involves technologies that are destabilizing of existing capabilities, and disruptive innovations entail disengagement with existing customers and secure income streams. There are compelling reasons why organizations try to avoid them.

This is where leadership comes in, providing the encouragement and resources to do things organizations find hard, but which are necessary for their continuing viability. Positive affirmation of the outcomes of innovation, through reviews and post-project assessments, and their wide communication throughout the organization builds support for new forms of learning. When positive results from innovation become remembered as organizational stories and corporate myths, they assist efforts to break with routine and institutionalized practices, and stimulate learning in all its forms.

Employment and work

There is a continuing debate on the impact of innovation on employment and its effect on the quantity and quality of jobs. Innovation has contributed to the massive historical shift in aggregate employment from agriculture to industry to the service sectors, but its impact on industries and organizations depends on their particular circumstances and choices.

The debate itself has a long history. Adam Smith would argue that increases in market size lead to greater opportunities for the division of labour, the replacement of people with machines, and potential deskilling. For Marx, automation inevitably led to labour replacement, wage reductions, and greater oppression of the workers. Schumpeter would argue that, as innovation both creates and destroys jobs, there will be mismatches between jobs and skills in declining industries and regions and emerging innovative new sectors, with painful adjustments needed and periods of skill shortages and unemployment.

One view has product and service innovation producing positive effects on jobs and skills, and process and operations innovation producing negative effects. As we shall see in Chapter 5, Edison created highly skilled jobs in his 'invention factory' and large numbers of unskilled jobs in his production factory. The skilled jobs were allied to product innovation, where thinking was at a premium; the unskilled jobs were linked to process innovation, where machinery reduced the need for thinking. But there is value in having skilled workers on production lines, and organizations often make choices about how they use innovations. The way machinery is designed and tasks configured affects the use of skills. Because of these choices, and as a result of the adjustments necessary as industries evolve in response to innovation, there are major incentives for individuals, employers, and governments to continually invest in education and training.

Organizations need to understand how innovation can be both personally rewarding and stressful; stimulating and frightening. It can provide incentives and motivation but also fear of change and loss of status. It can be divisive, with one part of the organization undertaking well-remunerated and satisfying work while others are poorly paid and dissatisfied. It may be exclusive, denying people access to jobs because of their particular education or gender.

Economic returns

Productivity, the index of outputs to inputs, increases when resources are used more efficiently. Increasing investment in labour and capital improves productivity ('more bucks'). It also increases when innovation, and improvements in technology and organization, contribute to what is known as multi-factor productivity (MFP) ('more bangs per buck'). Ultimately, economic wealth depends upon improved productivity, and this is frequently driven by innovation. MFP growth in the USA in the 1990s, for example, was linked to the information and communications industry and the use of its products in other sectors of the economy. Much recent growth in MFP has occurred in the service industries, such as retailing and wholesaling, and this can partly be attributed to the use of digital technologies.

Profitability is driven by a wide number of factors such as how much better and efficient organizations are compared to competitors at designing, making, and delivering things, and the preferences of customers for particular brand names and their preparedness to pay prices that provide the required return to innovators. Innovation contributes to profits by providing distinctive advantages in the sale of products and services; in their features, prices, delivery times, upgrade opportunities, or maintenance. Intellectual property can be sold and licensed, and new start-up businesses can be created, to produce profit from innovation. Large-scale innovative activity, in investments in R&D or plant and equipment, can deter competition and hence improve opportunities for profit.

For organizations to benefit financially from investments in innovation, they have to appropriate returns. In some circumstances, innovation can be protected by using intellectual property law for patents, copyrights, and trademarks. In others, protection derives from difficult-to-replicate skills and behaviours, such as the capacity to quickly stay ahead of competitors, being able to maintain secrecy, or retaining important staff. In all cases, the contributions of innovations to profits are frequently skewed, with most returns from a few innovations.

Technical standards that allow inter-operability between components and systems confer economic advantage. Those organizations that own standards, or whose offerings comply with them, have advantages over those that do not.

Continual pursuit: the case of IBM

The continual, broad-ranging, and challenging pursuit of innovation is seen throughout the history of the IBM Corporation. IBM is widely recognized as one of the most innovative companies in the world, playing a central role in the discovery and development of, among other things, supercomputers, semiconductors, and superconductivity. It has invested enormous resources in innovation. It spends billions of dollars on R&D each year, produces more patents than any other company, regularly creates iconic products and services, and its staff have won five Nobel Prizes. It has huge advantages in innovation compared to almost any other company in the world, yet its pursuit of innovation holds lessons for other organizations, not least that strategies need to continually evolve and you can never rest on your laurels.

IBM was incorporated in 1924, but its history can be traced back to Herman Hollerith's foundation of the Tabulating Machine Company in 1896. Hollerith (1860–1929) developed a machine using electricity and card processors to mechanize the data-processing of US Census data. He called the machine 'hardware' and the cards 'software'. Hollerith worked for a time at the US Census Bureau and was acutely aware of its need for improved efficiency in processing data. The 1880 Census had taken seven years to compile, and there were fears that the 1890 version was going to take longer. Hollerith's tabulating machine met the Census Bureau's requirement for fast and efficient data collection and management. By using it, the 1890 data were analysed in six months, saving millions of dollars, and it was subsequently used in censuses in Canada and Europe. By 1912, Hollerith had sold his business and, although he remained as chief consulting engineer, he had less and less association with the company. For many years, he had refused to respond to requests and ideas from the Census Bureau for improvements to his machines. When Hollerith's main patents expired in mid-1906, the Bureau developed a tabulator of its own, which it used in the 1910 Census. It took the arrival of Thomas Watson in 1914 to improve the technical performance of tabulating machines and improve the company's relationships with its customers.

As chairman of IBM, Thomas Watson (1874–1956) was instrumental in developing the company's use of electronics. He underwrote Harvard scientist Howard Aiken's 1930s research into developing a digital calculating machine. In 1945, in collaboration with Columbia University, he opened the first Watson Scientific Computing Laboratory, in New York. The IBM Thomas Watson laboratory remains today as one of the largest industrial research laboratories in the world. During the Second World War, the company developed very close relationships with the US government, especially in military ordnance and planning wartime logistics.

During his forty-two years at IBM, Watson built the company into a major international corporation. His son, Thomas Watson Jr, succeeded him as Chairman. From the late 1950s to the 1980s, following massive investments in R&D, IBM became the world leader in mainframe computers, especially with its System 360, launched in 1964 (see Figure 4). The System 360 remains in real terms one of the largest private investments ever made in R&D. The company, which was valued at $1 billion at the time, committed $5 billion to its development. Tom Watson Jr had 'bet the company' on its development. By 1985, IBM had 70 per cent of the world mainframe market. It had unmatched expertise in hardware and software, and its business skills made it one of the world's most admired companies.

4. The IBM system / 360 computer: IBM 'bet the company' on its development.

By the mid-1970s, the company began the pursuit for smaller computers. The IBM Personal Computer (PC), launched in 1981, was along with the System 360 one of the most iconic products of the last century; it essentially created the mass market for PCs. It emerged from an IBM development group that had failed in three previous attempts to create a PC. The successful development of the PC required the rejection of IBM's past strategy of self-reliance and developing everything itself internally. It decided to buy major components such as integrated circuits and operating software from small suppliers. Initially, the product was a huge success, capturing 40 per cent of the market.

By the late 1980s and early 1990s, however, IBM was in serious trouble and nearly went bankrupt. The IBM PC had helped sow the seeds of its own demise. IBM did not control the intellectual property rights to its components, and the small suppliers—Intel and Microsoft—grew quickly to become larger and more powerful than IBM, and supplied its competitors with their technology.

Furthermore, the overall culture of IBM remained focused on historically profitable mainframes at the same time as price competition from Japanese manufacturers led profit margins to collapse. The *New York Times* of 16 December 1992 was led to the opinion in an editorial that 'The IBM era is over...what was once one of the world's most vaunted high-tech companies has been reduced to the role of a follower, frequently responding slowly and ineffectively to the major technological forces shaping the industry.' The story of Herman Hollerith's rise and fall resonated again.

One response to IBM's 'near-death' experience was to appoint a new CEO, Lou Gerstner, the first appointed from outside IBM. The company went through a massive restructuring and fundamental shift in business strategy. It made the dramatic decision to sell its PC business to Lenovo in China, disposing of what was considered one of its core capabilities. It changed from being a supplier of technology to being a provider of solutions to customer problems. Its objective was to provide the best possible service for customers, even if this meant using competitors' technologies. At the same time, despite its financial difficulties, the decision was made that as the company's strength in the past derived from its 'science and engineering mindset', its research investment should continue in the future. The search was on to find increasing innovations from within the company's technological community and R&D centres. These internal sources essentially reinvented the mainframe using microprocessors and parallel architectures. IBM also became much more open to ideas from outside, attempting to break away from its past introspection and 'not-invented-here' syndrome. It began to use open technical standards and software rather than those it owned proprietarily and started to collaborate more in its technological developments, embarking annually on scores of collaborations with other organizations. Its new 'market-facing' innovations included supercomputing, ebusiness, social networks, and Web technologies.

The company presently makes extensive use of its intranet and social networking technologies to access and share ideas among its staff. With around 380,000 employees, half of whom are scientists and engineers, and twelve research laboratories globally, the company has massive technological skills on which to draw. IBM filed over 9,000 new US patents in 2017, more than any other company. Yet, as the company moves into new areas of technology, such as cognitive computing, and is competing in aggressive markets, such as cloud services, many uncertainties still confront it: emphasizing the challenges of remaining at the technological forefront, and that innovation is a journey not a destination.

Chapter 4

Stephanie Kwolek's new polymer: from labs to riches

Many people and organizations contribute to innovation. Large-scale surveys of innovative firms, such as the European Union's Community Innovation Survey, for example, show a wide range of contributors. These surveys also rank the importance of various sources, showing the most important of them lies within the organization. Innovation derives primarily from the energy, imagination, and local knowledge of employees identifying and solving problems. It is stimulated by innovative individuals and workplaces, and by formal organization structures and practices, such as R&D departments and management tools for developing new products.

Second in importance as sources of innovation, according to these surveys, are customers and clients, followed by suppliers of goods and services. Fairs and exhibitions, professional conferences and meetings, and academic and trade journals are reported to be important by a minority of firms. The least important sources, these surveys show, are universities and government research laboratories.

These rankings hide a much more complicated picture. Reliance on internally derived innovation, for example, makes organizations introspective and perhaps unprepared to deal with changes occurring externally in markets and technologies. Depending on customers for innovative ideas will likely produce conservative 'don't rock the boat' approaches. Universities are critically important contributors to invention for science-based sectors, and for innovative products and services at early stages of gestation, and they also educate and train employees with the skills to innovate.

As Josiah Wedgwood showed us, innovation usually involves the combination of ideas derived from many different starting places. The great scientist Linus Pauling said the best way to have a good idea is to have lots of them, and the same sentiment can apply to the pursuit of innovation from multiple contributors. Schumpeter's contention that innovation requires 'new combinations' among markets, technologies, and knowledge often entails integrating ideas from many different parts of the organization and with various external parties. The stimulus to innovate may not result from particular sources, with hierarchical contributions, but from multiple sources of ideas that intersect and blur in circumstances of intrinsic necessity and the impulsive quest for survival in volatile times.

Innovation is also affected by wider social, cultural, political, and economic factors. These include the contributions made by cities and regions, government policies, and the 'systems of innovation' organizations belong and contribute to. The case of IBM illustrates the diversity of sources it has used throughout its history in pursuit of innovation. We now turn to the various contributors to innovation.

Entrepreneurs and venture capitalists

In contrast with the large-scale activities of companies such as IBM, innovation also results from individual entrepreneurs using it to build new businesses. The term 'entrepreneur' began to be used in the early 18th century and is applied to individuals who discover, recognize, or create opportunities and then manage resources and bear risks to take advantage of them. Wedgwood demonstrates the substantial contribution entrepreneurs can make to innovation and economic development.

From Matthew Boulton in the 18th century, to Thomas Edison in the 19th, to Bill Gates in the 20th, and Sergey Brin and Larry Page in the 21st, entrepreneurs are commonly associated with the creation of technology-based companies. These companies grow rapidly on the basis of novel technologies that create new industries and transform old ones. Some entrepreneurs transform entire economies and societies. Boulton and his partner, James Watt, developed the steam engine and the world's first mechanized factory and helped usher in the Industrial Revolution. Edison, among many other contributions, developed electric power-generation technology and created the General Electric Company. Software from Gates's Microsoft popularized the personal computer; Brin and Page's Google transformed the use of the Internet; and both companies have changed the nature of work and leisure.

These are highly exceptional examples. Around half a million new companies are formed annually in the USA alone, and very few, if any, will be as successful as Microsoft and Google. Yet the creation of new firms, and the challenges they present to incumbent firms, is an essential element and a major contribution of capitalism. Schumpeter distinguished between two models of innovation. In the Mark I model, creative destruction is driven by the entrepreneurial task of 'breaking up old, and creating new tradition'.

Schumpeter's Mark II model recognized that entrepreneurship occurs in large established firms as well as newly created firms, reflecting the changing industrial realities as formally organized, large-scale R&D activities grew in scale from the 1920s. Entrepreneurship is therefore the organizational process by which opportunities are sought, developed, and exploited in many different kinds of company and organization.

In some circumstances, entrepreneurial start-up firms receive investments from venture capitalists who are prepared to assume higher risks than high street and investment banks. Many of the US success stories of entrepreneurial IT and biotechnology companies, such as Google and Genentech, received venture capital. Different international models of venture capital exist, but the US is often considered exemplary. US venture capital may include funds from private investors or corporations, and their managers may possess deep experience or knowledge of particular technological sectors and become engaged in the governance of start-up companies.

The objective of venture capitalists is usually to acquire shares in companies in their early years that then reap extraordinary returns after they exit when the firms have reached sufficient maturity to attract a purchaser or to be floated on a stock market. Among their portfolio of investments, venture capitalists recognize that the majority of returns will come from a limited number of cases. Generally, venture capitalists tend to invest in better-established, rather than new and speculative ventures, when technological and market opportunities have been clearly identified.

R&D

R&D is a significant, but not always essential, source of innovation. Investments in R&D help organizations search for and find new ideas and improve their capacities to absorb knowledge from external sources. R&D ranges from basic research driven by curiosity and little concern for its application, to highly practical problem-solving. Its expenditure reflects highly varied national, sectoral, and corporate commitments to its use in pursuit of innovation. Internationally, around \$2 trillion is spent annually on R&D according to some estimates. At an aggregate level, it is concentrated in a few main industries, including information and communication technology and pharmaceuticals. The USA is the major spender on absolute amounts of R&D. When relative expenditures on R&D are assessed—usually measured as a portion of a nation's gross domestic product—smaller European nations, such as Finland, Sweden, and Switzerland, head the list, committing over 3 per cent of GDP annually. A marked trend in recent years has been the rapid growth in R&D expenditures in Asian nations, such as South Korea, Taiwan, and China. Over 95 per cent of R&D globally is spent in the USA, Europe, and Asia (primarily northeast Asia), so many nations, especially in the southern hemisphere, cannot compete in this important source of wealth creation and growth.

There are wide differences among nations in the breakdown in R&D expenditures between that spent in business and government. In some countries, such as South Korea and Japan, business expenditure predominates. In others, such as Poland and Portugal, government is the major source of R&D spending.

In 1963, the Organisation for Economic Co-operation and Development (OECD) decided it would be useful for policy-making to have consistent international data on R&D statistics. Following a meeting in Frascati, Italy, this became known as the *Frascati Manual*. According to the *Manual* R&D comprises creative and systematic work undertaken in order to increase the stock of knowledge—including knowledge of humankind, culture, and society—and to devise new applications of available knowledge. It distinguishes between basic research, applied research, and experimental development. The *Frascati Manual* has been useful in building consistent data sets on R&D expenditures internationally. It has continually evolved and improved: the seventh edition of the *Manual* was produced in 2015. Nonetheless, significant problems remain in measuring collaborative R&D and activities undertaken in services. The OECD has also developed the *Oslo Manual* to guide national innovation surveys, the *Canberra Manual* for measuring human resources in science and technology, and a *Patent Manual* on the use of patent statistics.

Stephanie Kwolek's new polymer

Stephanie Kwolek (1923–2014; Figure 5) saved thousands of police officers and military personnel from death or disability. As a result of a traditional R&D process, she invented Kevlar, a fibre used in body armour. The product, which is one of the strongest fibres ever made, has over 200 applications, including in brake pads, spacecraft, sporting goods, fibreoptic cables, fireproof mattresses, storm protectors, and wind turbines. It has produced several hundreds of million dollars annually for the chemical company DuPont. It is best known, however, for its use in bullet-proof vests. In 1987, the International Association of Chiefs of Police and DuPont began a Kevlar Survivors' Club for those saved by the product from death or serious injury. It has over 3,000 members. Kevlar's protective properties have also been used extensively in the military.

Kwolek was born in New Kensington, Pennsylvania. Her steelworker father died when she was young, but she retained his curiosity: he had been a keen amateur naturalist. She remembers

5. **Stephanie Kwolek: inventor of Kevlar.**

spending hours designing and making clothes for her dolls and being very interested in fashion. She studied at a college that became part of Carnegie-Mellon University, and, unable to afford to study medicine, she majored in chemistry.

She decided she wanted to work for DuPont. DuPont was and remains one of the world's leading and most innovative companies. In the 1920s, it was one of the first companies to invest in basic research with the 'object of establishing or discovering new scientific facts'. It developed neoprene synthetic rubber in 1933 and nylon in 1938. With the shortages of male chemists resulting from the Second World War, women were being attracted into the chemical industry. During her interview, Kwolek forcefully demanded to know when she would hear about the job as she had another offer. The offer was made that evening.

Kwolek began work for DuPont in 1946. She worked at the DuPont Research Laboratory in Delaware for thirty-six years, having previously worked for the same group for four years in Buffalo, New York. Her job was to develop new polymers and ways of making them. Shortly after her arrival, she was given the job of looking for a breakthrough fibre to be used to make tyres lighter and stiffer. There was a concern at the time to improve vehicle performance to address a petrol shortage. Others had been offered the task, but weren't interested. Kwolek felt that, while her competence was recognized by male colleagues, she was often overlooked.

She liked the working atmosphere, however, and the challenges that arose, and as one of the few women scientists at the time, she worked exceptionally hard to retain her job after the men returned from the war. She was given a high degree of independence and freedom to do what she wanted. (She complained about modern research being too rushed and short-term with not enough time for thinking.)

Kwolek's specialization lay in low-temperature processes for the preparation of condensation polymers. In 1964, she discovered that the molecules of extended-chain aromatic polyamides formed under certain conditions a liquid-crystalline solution that could be spun into a strong fibre. She took her polymer, which was unpromisingly cloudy and thin, to a machine to be spun. She said the polymer had such strange features that anyone not thinking or being unaware would have thrown it out. The technician in charge of the spinner was deeply sceptical, thinking his machine would become bunged up by this contaminated substance, but was eventually persuaded to try. It spun successfully a product that was so strong that Kwolek had to repeat tests many times before she was convinced about her discovery. She did not tell anyone about her polymer until she was sure of its properties. Kevlar is heat-resistant, five times stronger than steel, and about half as light as fibreglass.

DuPont was immediately convinced of the value of Kwolek's new crystalline polymers, and the Pioneering Laboratory was given the job of finding commercial applications. She provided a small amount of fibre to a colleague experimenting with bullet-proof armour. Kevlar was introduced for this purpose in 1971. One of the reasons that it is used in such a broad range of applications is its versatility: it can be converted into yarn or thread, continuous filament yarn, fibrillated pulp, and sheeting. The new chemistry that Kwolek developed helped DuPont develop a range of other fibres, such as spandex Lycra and the heat-resistant Nomex.

Kwolek attributed her success to the way she could see things that others cannot. And she said:

> To invent, I draw upon my knowledge, intuition, creativity, experience, common sense, perseverance, flexibility, and hard work. I try to visualize that desired product, its properties, and means of achieving it...Some inventions result from unexpected events and the ability to recognize these and use them to advantage.

Kwolek had seventeen patents, including five for the Kevlar prototype. She won numerous prestigious awards and spoke about the great need for recognition of scientists and other people who do things that benefit mankind. She admitted to being very pleased when a police officer asked for her autograph on the jacket that saved his life.

The case of Kwolek and Kevlar epitomizes the corporate R&D department's contribution to innovation. It also points to some of its shortcomings. The polymer was eighteen years in development, and its commercialization took seven years. Few, if any, organizations have the capacity to take such a long-term approach nowadays.

Graphene is among the most important new materials to emerge from basic research in recent times, and is beginning to be used in many different industries and applications. First discovered in 2004, after a decade of applications-focused R&D it is now found in plastic electronics, clothing, and water decontamination systems. It is being used to make bicycle tyres that have 10 per cent lower rolling resistance, and better traction and puncture resistance, compared with traditional tyres. It may in future make aircraft and cars lighter and stronger and therefore more efficient, but as with Kevlar finding applications of basic research can take many years.

Customers and suppliers

Innovations do not succeed unless customers or clients use them, and if the users of new products and services are involved in designing what they need, there is generally a better chance of success than if something is being designed for them. Demands and needs can never be articulated fully and communicated completely across organizational boundaries between producers of innovation and their customers and suppliers, and active engagement between them overcomes these barriers.

In some fields, such as medical instruments, the innovator is commonly the user of the innovation. Surgeons and medical practitioners are regular contributors of ideas for new tools and techniques that help them do their jobs better. The world's largest producer of implanted hearing devices, Cochlear, originated with Professor Graeme Clark, a medical researcher whose father was profoundly deaf. Clark was highly sensitized to the suffering of people whose deafness could not be helped by hearing aids, and was driven to improve their lives.

According to one estimate, up to one-quarter of all men over 30 suffer from sleep apnoea, a condition that causes potentially dangerous breathing irregularities when they are asleep. Respiratory medical devices can help manage the problem. The origins of the world's largest manufacturer of respiratory devices—ResMed—lay with Professor Colin Sullivan, a medical researcher working in a hospital sleep clinic. He overcame the problem by regularly blowing puffs of air up the nasal passage. Fortunately for sufferers and their partners, as a result of continuous design improvements the present discreet and quiet devices are a big improvement from the original version built from a gas mask and vacuum cleaner.

Some companies go to great lengths to engage customers in designing new products. When Boeing developed the 777 aircraft, it involved its major customers, United, British Airways, Singapore Airlines, and Qantas, in trying to understand the demands of the market. It needed to know about optimum passenger loadings for airlines' favoured routes. But it also worked to understand the demands of the users of the aircraft: the pilots and aircrew, maintenance engineers, and cleaners. It aimed to be sympathetic to flight attendants brewing coffee in turbulence and maintenance engineers fixing an external component in Alaska at midnight in minus 40 degrees centigrade or Jeddah at midday in 50 degrees. As Boeing developed the 787, it built a website to gain immediate input into the design process from interested parties around the world. Around 500,000 people voted on the choice of the aircraft's name: Dreamliner.

Software companies sometimes release their products in 'beta' form, that is, in prototype, to allow users to play with the software and suggest improvements. Essentially, customers do much of the final polishing of the product. This strategy is pursued for products that are proprietary, when the company aims to profit from them. This is different from open-source software, such as the web browser Mozilla Firefox and the operating system Linux, which is built, maintained, and continually improved by networks of volunteer programmers.

Excluding customers from the process of product improvement can be very shortsighted. When Sony developed its robot dog—the Aibo—it kept its software code secret. A community of hackers grew up, developing a much more extensive range of moves for the robot, including a number of amusing dances that made it a much more attractive product for customers. Sony sued the hackers and closed down their community, but soon after recognized its mistake and realized that the company could learn from the software developed externally. Sony no longer produces the Aibo, but its subsequent products have benefited from the technology developed for the robot dog in areas such as visualization.

Customers can also inhibit innovation. They can be conservative and complacent and locked into ways of doing things that preclude novelty and risk. Clayton Christensen identified the 'innovator's dilemma'; the problem of listening too closely to customers. If innovators only respond to the immediate demands of customers, they often miss big changes occurring in technologies and markets that may eventually put them out of business. Here there is advantage in working with 'lead customers', governments, firms, or individuals that are prepared to take risks to promote innovation in the belief that greater benefits will accrue than pursuing the safer short-term option of not innovating. In the 1980s, Roy

Rothwell described the relationship between Boeing and Rolls Royce as 'tough customers; good designs', implying that Boeing's very demanding requirements of its aero engine suppliers made Rolls Royce design and produce better products.

Innovative suppliers are also major stimulants to new ideas. In the automotive industry, a high percentage of the value of a car is bought from suppliers of components, accounting in Toyota's case for up to 70 per cent of the car's total cost. Toyota enjoys very close relationships with Nippondenso, a very large components supplier of innovative products such as lighting and braking systems. The automotive supplier Robert Bosch plays a similar role in the European car industry.

Large automotive companies use numerous methods, including websites and technical conferences and fairs, to encourage their suppliers to provide innovative solutions to the problems they face. Innovative automobiles are based on suppliers of innovative components to car companies. The task of the car manufacturer—or the organization responsible for integrating any system of different elements—is to encourage innovation in suppliers of modules or components while ensuring the compatibility of components with overall design architectures or systems.

The encouragement of innovative suppliers is also a key objective of many governments. In the USA, the Small Business Innovation Research scheme uses the government's enormous procurement budget to support small companies by purchasing innovative products and services. This particular government scheme invests more in innovation in early stage start-up companies than the US venture capital industry.

Collaborators

Innovation rarely results from the activities of single organizations and more commonly occurs when two or more organizations collaborate. For many organizations, the benefits of using collaboration to contribute to innovation outweigh the costs of sharing the returns to that innovation. Collaborations take the form of joint ventures and various types of partnerships, alliances, and contracts that involve joint commitments to mutually agreed aims. They can be with customers and suppliers, organizations in other industries, and even with competitors. They are a feature of the world's industrialized economies, with some collaborations operating for many decades.

Organizations collaborate to reduce the costs of developing innovation, to access different knowledge sets and skills to the ones they possess, and use it as an opportunity to learn from partners about new technologies, organizational practices, and strategies. In uncertain and evolving circumstances, innovating collaboratively provides a greater chance of success than going it alone. Information, communications, and other technologies have made collaboration cheaper and easier. Governments around the world have actively promoted collaboration as a source of innovation. And organizations have become less self-reliant and more open to collaboration in their strategies for innovation.

Different types of collaboration work better in different situations. When the objectives of collaboration are clear, or the focus is on quickly reducing costs, it works better when organizations are similar. The opportunities for misunderstandings and miscommunication are fewer. When objectives are emergent, and the objective is exploration and learning, collaboration benefits from dissimilar organizations working together. More is learned from variety than uniformity. Larger numbers of partners increase the scale of effort; fewer partners improve speed.

Collaboration can be difficult to manage. Partners may have different priorities and organizational cultures. There are many opportunities for misunderstanding, as the following, perhaps apocryphal, anecdote reveals. Some years ago, a collaboration was proposed between a group of IBM and Apple staff. Prior to the first joint meeting, IBM staff discussed their approach. Conscious of their reputation for formality—blue suits were the uniform of the day—they decided to put the Apple staff—who were usually casually dressed—at ease by wearing their weekend clothes to the meeting. They arrived in jeans and sweatshirts to find the Apple staff sitting uncomfortably in newly purchased blue suits. That this could occur among organizations in the same industry and country highlights the potential problems that might occur in collaborations in different sectors and nations.

Universities

Clark Kerr, renowned social scientist and president of the University of California, was highly prescient in identifying the importance of universities for economic development when he wrote in 1963 that:

> the university's invisible product, knowledge, may be the most powerful single element in our culture...the university is being called on to produce knowledge as never before...And it is also being called upon to transmit knowledge to an unprecedented proportion of the population.

He argued new knowledge is the most important factor in economic growth and highlighted the role of the university in developing new industries and generating regional growth, and emphasized the contribution of the entrepreneurial professor, consulting and working closely with business. In the intervening decades, universities have been increasingly encouraged by governments and business to devote their energies to actively translating their knowledge into economic activity, a policy they have often enthusiastically endorsed. This activity is now so elevated that it has emerged, according to some, to be as important a university function as research and teaching. The ways knowledge is transferred to industry, and universities contribute to innovation, are often conceived too simply, however, and the paths to market are commonly complex, multifaceted, and subtle. The idea that ideas and knowledge are something that universities produce and 'transmit' to industry has also been replaced by the notion that it is co-created and exchanged.

Teaching. By educating skilled undergraduates, graduates, and post-doctoral students, universities prepare a labour force's capacity to create and apply new ideas. The history of the successful development of new industries—such as electrical, chemical, aeronautical, and IT—is in major part explained by the provision by universities of sufficient numbers of graduates with the requisite new skills, especially in engineering and management. It is said that the best form of knowledge exchange between university and industry is carried on two legs and by the movement of problem solvers from university to industry.

It is not only science and engineering graduates who contribute to innovation. At various times, philosophers and anthropologists have been in demand in Silicon Valley, and the creative industries provide a home for many humanities students. Business schools increasingly provide courses in innovation management and entrepreneurship for students of all disciplines. Some in the management field discuss how successful firms are said to require a combination of 'I'-shaped people, with deep knowledge in a particular field, and those that are 'T'-shaped, having a breadth of knowledge with a particular specialism. The capacity to see connections 'across the T' between various disciplines is a major stimulus to innovation, but poses significant challenges for educators who as well as teaching about areas of knowledge have also to teach about the connections between them.

Technical colleges also play an important role in innovation, for example in training technicians to produce prototypes and instrumentation which they themselves occasionally commercialize.

Science and research. Science, from the Latin *scientia*—knowledge—has been a feature of human development since the first civilizations. The application of science in industrial innovation, however, only began in earnest during the Industrial Revolution and has been most particularly a feature of the last 150 years or so.

One of the traditional distinctions in research, seen in the *Frascati Manual*, is between that which is 'basic' and that which is 'applied'. The former is thought to be curiosity-driven, with no consideration of its application, and is the particular concern of universities. The latter is believed to be directed towards an identified use, and is usually explored in industry. Yet some businesses invest substantially in basic research and universities conduct extensive applied research, especially in professional departments such as medicine and engineering.

Furthermore, as Donald Stokes argued, the classic distinction between 'pure', basic research, driven by a desire to understand, and applied research, with the purpose to be used, fails to capture a third category which aims to do both by improving understanding and being useful. He calls this 'Pasteur's Quadrant' of use-inspired basic research (see Figure 6). Pasteur's microbiology research was always concerned with useful applications, but it also created a new field of scientific understanding. Stokes contrasts this with Bohr's research in physics, in which his understanding of atomic structure provided a basis for developing the theory of quantum mechanics, and Edison's research that was driven by a concern for use and profit, although he was also influenced by theory. There is a direct and obvious connection between research and innovation in Edison's and Pasteur's Quadrants: the connection in Bohr's may or may not occur, and if it materializes, it might be in unexpected or unimagined areas. Bohr, one imagines, would have had little appreciation of how quantum theory is used to explain lasers and potentially provide the basis of future quantum computers

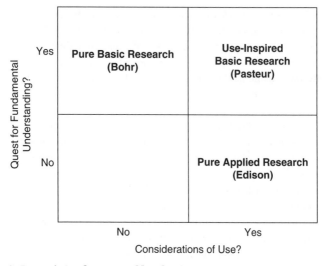

6. **Pasteur's Quadrant, Donald Stokes (1997).**

that make use of the quantum states of subatomic particles to store information.

Similarly, in their short letter to *Nature* on 25 April 1953, Watson and Crick modestly stated: 'We wish to suggest a structure for the salt of deoxyribose nucleic acid (D.N.A.) This structure has novel features which are of considerable biological interest.' They did not imagine the considerable commercial interest that was to emerge more than twenty years later, or the way in which their discovery has transformed old and created new businesses with the development of biotechnology and genomics.

In reality, basic and applied research are elements of a continuum, with many interconnections. Applied research may result from basic research findings, and basic research may be undertaken to explain how an existing technology works. One of the most useful outcomes from pure basic research is the instrumentation developed to assist experimentation. The computer, laser, and Worldwide Web were developed to this end, with little appreciation of potential industrial use as now ubiquitous innovations.

When we consider the world's most complex scientific and social questions, including global warming, sustainable energy, food security, and genetic engineering, the answers will depend on fundamental understanding developed by universities and its practical use in industry.

Engagement. Dr Jonas Salk was reputedly once asked who owned the polio vaccine he had developed. His answer was, 'why the people, I would say'. Such a response is unlikely today. Since the passing of the Bayh-Dole Act in the USA in 1980, which allowed research institutions to own the results of publicly funded research, universities in developed economies have become preoccupied with making money from their research. This has usually taken the form of patent-protected intellectual property, licensed to businesses, or through start-up companies, spun out of and part-owned by the university. Evidence suggests, however, that the number of successful instances of this model of commercialization is limited. There are some impressive success stories, such as the biotechnology firm Genentech. This company was formed in 1976 to help commercialize the discovery of recombinant DNA at Stanford University, and was sold to a Swiss pharmaceutical company in 2009 for just under $50 billion. Such companies, however, are a tiny fraction of the total amount of entrepreneurial activity encouraged by universities.

The focus of the majority of attention by governments, and indeed by many universities themselves, has been on issues such as patents and licensing, contract and cooperative research, and incubation and entrepreneurship centres. Also important are the social and networking activities that are crucial to the 'conversations' between universities and business about new developments and their potential applications. Although for many businesses, particularly smaller ones, the purpose of collaborating with universities is immediate problem-solving, larger firms will engage in broader dialogue with universities to learn about the directions of future research. Businesses claim that the attraction of working with universities is that they have different cultures to their own. University staff have more time to think about and test new ideas.

As contributors to inventive ideas and knowledge creation and diffusion, universities and research institutes need continually to communicate their capabilities and assess how they should best engage with external parties. Governments, businesses, and philanthropists cannot be expected to invest in universities and research institutions as providers of innovation without them fully articulating their broad contributing roles.

Regions and cities

Innovation agglomerates by localizing within particular geographies, such as in the Staffordshire Potteries. It does so for economic reasons, as proximity reduces the costs of transactions and transportation, and firms in close association stimulate the creation and diffusion of innovation through improved awareness and knowledge of each other. Innovation clusters for social and cultural reasons, including the advantages derived from shared identity and higher trust in affiliated and cohesive groups. Communications are assisted by propinquity because knowledge is sticky and travels badly from its source, especially when it is complex or tacit and cannot be written down.

The best-known innovative region is Silicon Valley, near San Francisco, an area of high technology business concentration and employment, and one that has stimulated countless and often fruitless attempts at replication around the world. A number of factors contributed to the development and growth of Silicon Valley. Government played a central role, from the gift of land to local universities to stimulate industrial development, to being a large-scale customer of high-tech goods in defence markets. Universities have contributed their research and the education and training of scientists, technologists, and entrepreneurs. Institutions such as Stanford University have proactively developed policies to encourage academic engagement with business in fields such as electronics and IT. Numerous high-tech businesses have started up and some have grown rapidly into major corporations, such as Hewlett-Packard, Apple, and Intel, assisted by a highly skilled and mobile labour market attractive to talented employees, links to university research, and ready access to professional services such as venture capitalists and patent lawyers. These factors contribute to a local culture, or 'buzz', that is technology-focused, risk-taking, and highly competitive, and it creates a virtuous circle of initiative and reward. It has created enormous wealth and extensive experience of innovation and entrepreneurship to be reinvested back into new initiatives.

It is often not regions but cities that provide the locus of innovation. Throughout history, cities have at various stages been associated with creativity and innovation, from Athens in the 5th century BC to Florence in the 14th century, to fin de siècle 19th-century Paris.

Cities are major contributors to the supply and demand of innovation. Most patents emanate from and R&D is conducted in cities, and their higher disposable incomes ensure greater consumption of innovation. Some cities are renowned as centres of learning, such as Oxford or Heidelberg; others for their engineering ingenuity, such as Stuttgart or Birmingham; financial and services innovation, such as London and New York; and creativity and design, such as Copenhagen and Milan. Some cities are known for their particular technology expertise, such as Bangalore and Hyderabad in India; or their support for technological entrepreneurship, such the Hsinchu area in Taiwan or the Jiangsu, Shenzen, and Zhongguancun areas in China. The efforts of many city governments have been directed towards policies to identify and harness innovation that provides comparative advantage over other cities internationally. Although many have been blinded by the attraction of the Silicon Valley, technology-led, model, it is important that others have different approaches by, for example, addressing health, fashion, or the media. The issues of innovation in cities will be discussed further in Chapter 6.

Government

The debate about the role of the government in supporting innovation commonly reflects political ideology. State intervention in innovation is considered essential in many nations, including most Asian countries who see it as crucial to economic and social development. However, in more 'free-market' economies, such as the USA, rhetorically at least, government intervention is regarded sceptically and avoided, usually with reference to government's inability to 'pick winners'. Nonetheless, the past polarities of the views that argue, on the one hand, that interventionist innovation policies distort markets and promote inefficiencies or, on the other, are essential components of sound economic planning and effective industry policies are tending nowadays towards a pragmatic middle ground. Here it is recognized that government has an important role to play in innovation, but policies have to be selective.

Governments contribute to innovation in many ways apart from their innovation policies. A stable, growing economy enhances the preparedness of firms and individuals to invest in innovation and take risks. Effective monetary and fiscal policies are crucial in providing confidence in the future. A nation with more wealthy firms and individuals is better placed to be innovative. Good educational policies produce employees and entrepreneurs with skills to create, assess, and realize opportunities for innovation. Well-educated citizens are more capable of contributing to national debates on innovation, and determining which sciences and technologies are acceptable, and what form new products and services should take. Government investments in research—which in developed nations account on average for about one-third of total expenditures on R&D—provide many of the opportunities for innovation. These investments can often take a longer perspective than those made in the private sector. Competition policies prevent monopolies erecting barriers to innovation; trade policies increase the size of markets for innovative products and services; intellectual property laws can provide incentives to innovate; regulations in areas such as environmental protection stimulate the pursuit of innovation. Free and open access to information stored by government increases opportunities for innovation. Innovation in a highly digitally connected world is inhibited unless government acts to ensure personal privacy and encourage ethical codes of practice when it comes to the collection and use of data. Open immigration policies allow the flow of talent from overseas and are the source of diversity, which is so important for innovative thinking. Industrial relations laws can help provide equitable, secure, and participative workplaces encouraging of innovation.

Governments can encourage innovation through their procurement power: they are the primary purchaser of innovation in any nation. Public expenditures on IT, infrastructure, pharmaceuticals, and many other areas, exceeds that of the private sector, so government purchasing is a major stimulus to innovation.

Leadership by governments can set a tone or atmosphere in which innovation is encouraged. When the political discourse is future orientated and ambitious—think of John F. Kennedy's plan to have 'man on the moon', or Harold Wilson's 'white heat' revolution of science and technology, or President Xi's vision and mandate for innovation to modernize the Chinese economy—it is more supportive of innovation than when it is relaxed and comfortable with the status quo or attached to a romanticized vision of the past. Public servants are more likely to support innovation when they are not fearful of censure for the slightest mistake or risk-taking behaviour.

Apart from these forms of support, many governments develop specific innovation policies. These have tended in the past, especially in scale of expenditures, to focus on R&D, usually in the form of tax credits: by spending on R&D, firms can reduce their tax bills. There has been a plethora of other types of policy designed to encourage innovation. These include demonstration schemes, highlighting the benefits of particular innovations; consultancy schemes, helping organizations to improve their ability to innovate; investment schemes offering subsidies or increasing the amount of available venture capital for innovation; and creating new intermediary organizations that help build connections between research and business.

Many justifications for government innovation policy have been championed. These include, at their most practical, fear of international competition. The US government's response to the growing dominance of Japanese competition in semiconductors in the 1980s, for example, led it to create a well-funded consortia of US manufacturers, Sematech, directed to produce competitive technologies. Many pan-European schemes in the IT industry during the same period were designed to build the capability in Europe to resist US and Japanese competition. Some policies purported to encourage innovation are simple forms of industry support—or corporate welfare, to take a less charitable perspective. Schemes around the world to continually support the ailing car manufacturing industry in marginal electoral constituencies would provide a case in point.

Much of the justification for government intervention is presented in the form of an argument about 'market failure'. R&D, it is argued, produces knowledge that can be cheaply accessed by the competitors of those who undertake the risk of investing in it. The 'public' returns to investments thereby exceed the 'private' returns, and therefore there is a tendency towards under-investment. To address this market failure, government justifies the financial support of R&D in firms.

This form of support, which assumes the bulk of government investment in innovation policy, has several limitations. First, it is concerned with R&D, which is but one input into innovation, and in many industries and circumstances not the most important one. What is construed as 'R&D' might also be limited, excluding important inputs to innovation such as software development and prototyping. Second, it misunderstands the investments required for public returns. The capacity of firms to access the R&D conducted by others is not costless. It requires investments to allow recipients to absorb the new ideas. Third, if market failure leads to sub-optimal investments in R&D, then there must be an optimal level, but there is little evidence about what this might be. Fourth, the delivery mechanisms of R&D support are highly generic, usually taking the form of tax credits for spending on R&D, rather than its performance. There is rarely provision for support of additional R&D to that which would be invested in without government money. The tax allowances are broadly available across industry, without the capacity to select strategic targets. Furthermore, application and compliance costs are usually resource-intensive, favouring larger, wealthier applicants rather than their generally more deserving smaller counterparts.

An additional case for government innovation policy can be made from the perspective of systems failure. Despite reservations about the dangers of the mechanical and predictable ways the national innovation systems are viewed by government, as opposed to their commonly fluid and unpredictable reality, there is value in conceiving them from a government perspective. Government is the only actor capable of taking an overall national innovation systems view, and the only one able to influence its whole construction and function. It can assess performance, identify gaps and weaknesses, and support institutions and policies that build connections. The challenge for policy-making related to national innovation systems is that much attention is directed towards describing the components of the system, rather than what the system does, or perhaps even more importantly, what it should do.

The essential criterion for innovation policy is the extent to which it encourages and facilitates the flow of ideas across the economy and within national innovation systems, and enhances the chance of their being successfully combined together and implemented. These flows of ideas occur in many, frequently unpredictable, directions: between the manufacturing, service, and resource industries; the public and private sectors; science, research, and business; and internationally within research networks or production supply chains. Innovation policy should therefore be concerned with the encouragement of the flow of ideas, the capacity of organizations to receive and use them, and the impediments which prevent effective connections between the various contributors to innovation.

Encouragement of the flow of ideas comes from open access to information and publicly funded research results, institutions that 'broker' connections between users and suppliers of knowledge, regulations that stimulate or at least fail to impede innovative investments, and judicious intellectual property laws that deal with the profound challenge of providing the confidence in their ownership to encourage trading, without providing the disincentives that occur from the award of monopoly positions. The receptivity to innovation in organizations depends upon the skills, organization, and quality of management of the recipients. Blunt policy initiatives such as R&D tax concessions are valuable only to the extent that they increase the quality and quantity of organization's capacity to select and use new ideas.

Systems

The incredible success of Japanese industry in the 1970s and 1980s led to a search for its explanation. One analysis argued that it resulted from Japan's ability to organize the various elements of its economy into a national innovation system. In this view, the Japanese government played a central role coordinating large corporations' investments in important and emerging areas of industrial technology. Japan's strength in consumer electronics, for example, was believed to have resulted from the country's highly effective Ministry of International Trade and Industry (MITI) collecting information from around the world on new technologies and organizing the efforts of large electronics firms, such as Toshiba and Matsushita, to take advantage of new opportunities. The ability of the Japanese government to do this was exaggerated, but it did play an influential role and researchers began to think about the contributions to innovation made by national institutions and characteristics, and the ways they combined into a system. The search was on to try to understand the role of the major players and the most important of their interactions, and provide some capacity to encourage innovation at national level.

Early research into national innovation systems took two forms. One, primarily US in focus, took an economic and legal perspective, and concentrated on the nation's key institutions, including those of research, education, finance, and law. The characteristics of effective national innovation systems were seen to be high-quality research, providing new options for business; education systems that produced well-qualified graduates and technicians; the availability of capital for investments in risky projects and new and growing ventures; and strong legal protection of intellectual property. The other approach, primarily Scandinavian in focus, was concerned more with the quality of business relationships in a society. The characteristics of effective national innovation systems were seen to be close ties between customers and suppliers of innovation, influenced by the amount of trust between people and organizations in a society and the learning this engenders.

These approaches were initially developed by academics interested in analysing and understanding the reasons why innovation occurs, and why it takes particular forms. The question emerged, for example, of why some nations, such as the USA, are particularly strong in radical innovation—explained by its strength in basic research—and why others, such as Japan, are very strong at incremental innovation—explained by efficient coordination of information exchange between customers and suppliers. The idea of national innovation systems, however, quickly took hold in government and public policy circles, as a way of prescribing and planning how institutions and their relationships could be configured. International organizations, such as the OECD, have produced numerous reports on the institutions of various nations, but these tend to be highly descriptive and static, failing to explain how national systems evolve over time. They do make the valuable observation, however, that what matters is not only the institutions that exist in a nation, but how effectively they work together.

At the same time that research on national innovation systems was blossoming, some began to ask whether the nation was the most useful level of analysis. The question was raised of why nations are often successfully innovative in some industries and regions, but not in others. The USA has Silicon Valley in California, but it also has the Rust Belt of declining heavy engineering and steel industries in the northeast. Researchers have argued the importance of regional, sectoral, and technological innovation systems. They examine the characteristics of successful regions, such as Route 128 around Boston and Cambridge in Massachusetts, Cambridge in the UK, Grenoble in France, and Daejon in South Korea. They examine differences in patterns of innovation in the machine tool and textile industries. And they explore why innovation in biotechnology occurs differently from that in nanotechnology. Given the high levels of investment in innovation made by large multi-national companies, operating across borders, researchers have also argued the role of global innovation systems.

The notion of systems of innovation is a helpful framework, but social systems are not engineering systems for which components and their interactions are known, planned, and constructed. Unpredictable events occur, and the systems evolve and change in unexpected ways. The early lead in biotechnology research in Harvard University, for example, was lost to Stanford University because of the election of a populist mayor in Boston who built on people's fear of unknown consequences of genetics research. What matters is thinking about the ways in which all the innovation-supporting institutions interrelate and evolve over time along with business practices and relationships. And whatever the level of analysis—global, national, regional, sectoral, technological—what is important is to understand how they relate to one another and co-evolve.

Governments have a major role to play in developing innovation systems, as seen especially clearly in Asia. The industrialization of Asia in recent decades has led to the extraordinary social and economic development of the region. South Korea, for example, has been transformed from the second poorest country on Earth in the 1950s to a member of the OECD, the group of the world's thirty richest nations. Asian industrialization has required rapid developments in research, education, finance, and law. Countries such as India, South Korea, Taiwan, and Singapore have developed coherent national innovation systems and become important international contributors to innovation. The models of development have varied. South Korea, for example, has depended on large conglomerate companies, Taiwan on networks of small firms, Singapore on direct foreign investment by large multinationals, and China has pragmatically used all of these approaches. In East Asia, the process of development has been strongly directed by the state, and this is, of course, especially so in China.

China has experienced the most rapid and remarkable industrial development in history. From the devastation of the Second World War, civil war, and cultural revolution, it has emerged as a global manufacturing powerhouse, investing massively in science, technology, and education, and potentially challenges Western hegemony in innovation. The transformation of innovation in China has resulted from strong political leadership. President Hu Jintao called for an innovation-oriented country, pursuing a path of innovation with Chinese characteristics, a theme further promoted by President Xi Jinping. The political discourse in China refers to 'harmonious growth', and the imperative for inclusive development is the most important challenge confronting innovation in China. This encompasses the need to use innovation as a means of reducing income disparities between the poor and wealthy, and the economic disparities between coastal regions and inner China. The evolution of China's national innovation system to one that allows it to compete equally with the West in innovation is incomplete and continuing, but it is clear that the state will continue to play a strongly directive role. Like most things in China, when decisions are made they occur at scale, so when, for example, it was decided to corporatize government industrial research institutes, this involved around 2,000 organizations employing around one million people, and their transformation occurred rapidly.

Chapter 5
Thomas Edison's organizational genius

Organizations have choices on how they organize themselves for the continually evolving challenges in innovation; the structures and procedures they adopt, the staffing and incentives they use. These reflect their strategies and innovation objectives.

Edison

Thomas Edison (1847–1931) is remembered for his inventiveness and the wide range of innovations he introduced. He held over 1,000 patents, and, among other remarkable achievements, he developed the phonograph, electric light bulb, and electrical power distribution, and improved the telephone, telegraph, and motion picture technology. He founded numerous companies, including General Electric. He was also responsible for pioneering a highly structured way of organizing innovation, and it is this that is our concern here.

Like Josiah Wedgwood, Edison was the youngest of a large family of modest circumstances, received little formal education, started work early, aged 12, and was afflicted by a disability, deafness, that influenced his life and work. He was similarly driven and hard-working, and shared with Wedgwood an appreciation of Thomas Paine, which also influenced his democratic worldview.

Edison could be blunt, irascible, and impatient, but he could also be personable, kind, and generous.

Edison began his working life as a telegraph operator, and started to experiment during the night shift, when he could not be observed. His first patent, an electric vote recorder, was awarded when he was 22. The notoriety of his inventions moved him from his humble beginnings into high circles. He demonstrated the phonograph to President Hayes in the White House in 1878, and was close friends with Henry Ford. He is reputed to have influenced Ford over the potential of petrol engines. His business partners included the leading capitalists of the day, such as J. P. Morgan and the Vanderbilts.

Edison's approach to business was relentless and ruthless. He demanded continual improvement in innovations from his employees and energetically denigrated opposition. His campaign against alternating current (AC) and promotion of direct current (DC), his preferred option for electricity transmission, sank to the unsavoury level of a publicity war over their relative merits for the electric chair. Edison did not shrink from demonstrations of electrocuting animals with AC to reveal its dangers. These included the unfortunate, if ill-tempered, Topsy the elephant, whose demise at Lunar Park was filmed by Edison for further publicity value. AC, the superior system, eventually became dominant, and the cut-throat nature of the battle between these competing technical standards clearly shows the value of owning the dominant version.

While Edison enjoyed great commercial successes, he had his fair share of failures. There were comparatively expensive and unproductive diversions into mining and the manufacture of concrete. He failed to recognize public interest in the celebrity of musicians, when for years he refused to name them on recordings. With characteristic aplomb he claimed never to have failed, but to have discovered 10,000 ways that didn't work.

Possession of intellectual property was crucial for Edison. Patents emerging from research in his laboratories were attributed to Edison irrespective of his contribution. One of his long-term assistants said: 'Edison is in reality a collective noun and refers to the work of many men.' Fiercely protective of his own patents, he occasionally disregarded the intellectual property of others. He and his business partners regularly used patents to block the development of competitors.

Feted during his lifetime, and called a 'wizard' by the press, he faced hostile criticism from rivals. Critics included Nikola Tesla, who had every reason to be bitter. Tesla was working for Edison when he developed AC, prior to commercializing it with the Westinghouse Corporation. Tesla claimed he was not paid what he was promised. In later life, Edison regretted the way he had treated him. It is speculated that the reason Edison did not proceed with AC himself, despite having many opportunities to do so, was that he had not developed it himself; a case of the 'not-invented-here' syndrome. After Edison's death, Tesla reported for posterity his ex-boss's utter disregard of the most elemental rules of hygiene.

The way Edison organized his inventive efforts derived from his overall approach to innovation. He always pursued several lines of research, wishing to keep options open until the strongest contender emerged, at which point resources and effort would be concentrated. By working on numerous projects simultaneously, Edison hedged his bets so future income streams did not depend upon one development. He was well aware of how pursuing one problem would lead to others, often completely unexpected, and he understood the value of chance, serendipity, and 'accident'.

Edison explored how ideas from different areas of research could potentially be combined, and he had a strategy of re-using proven components of other machines and applying them as building blocks in new designs. Edison said he readily absorbed ideas from every source, frequently starting where others left off. The development and commercialization of the light bulb, for example, combined ideas by drawing on a network of researchers, financiers, suppliers, and distributors. Although the idea of the light bulb had existed for decades, Edison, by using low-current electricity, a carbonized filament, and a high-quality vacuum, developed a product that was relatively long-lasting. His principles were to experiment and prototype as much as possible on a small scale, and to make designs as simple as possible. Once a breakthrough had occurred, he appreciated it would take a great amount of continuing research and experimentation to turn it into a successful product. He said it usually took him five to seven years to perfect a thing, and some things remained unresolved after twenty-five years. As he said: 'genius is one percent inspiration, ninety-nine percent perspiration'.

Edison understood that most value returned to the controller of the technical system, not the producer of its individual components, who was dependent upon the system configuration. His systems thinking was most apparent in the development of the electricity distribution industry that began operating in New York in 1882. Recognizing people's apprehension of the unfamiliar, Edison cleverly mixed the new and the existing in his electricity system. He used recognizable infrastructure to deliver electricity, including putting wires underground like gas pipes and utilizing existing gas fittings in homes.

Like many of his innovations, Edison's approach to organizing his research laboratories built on the experience of others. The telegraph industry in which Edison began his career had a number of small research shops with ranges of experimental equipment. Edison had conducted experiments in one such shop in Boston, and upon his arrival in New York in 1869, he used another before setting up his own laboratory in Newark to make his design of stock ticker machines.

Edison's organizational innovation lay in the range and scale of the research activities undertaken. He invested more financial and technological resources in the search for innovation than any other organization had done previously. Edison established the Menlo Park laboratory in 1876 so he could devote himself entirely to the 'invention business'. Located 25 miles from Manhattan, in what was then a small village, by 1880, seventy-five of the 200 Menlo Park residents worked for Edison. Menlo Park began with an office, laboratory, and machine shop. Over the years, Edison added a glass house, photographer's studio, carpenter's shop, carbon production shed, blacksmith's, and additional machine shop. He also added a library.

At this time, only a few of the best universities in the USA had laboratories, and these were ill-equipped and focused mainly on teaching. Yet Edison had fine scientific equipment, including an expensive reflecting galvanometer, electrometer, and photometrical devices. Within a couple of years, his stock of tools was worth $40,000 ($900,000 at 2016 prices—a massive investment at the time). Edison's objective was to have all tools, machines, materials, and skills needed for invention and innovation in one place.

At its peak, Edison had more than 200 machinists, scientists, craftsmen (only men being employed in this work, at that time), and labourers assisting with inventions. They were organized into teams of ten to twenty, each working simultaneously in transforming ideas into working prototypes. As everyone in the team had the same objective, communications and mutual understanding were at a premium. In six years at Menlo Park, Edison registered 400 patents. He aimed for a minor invention every ten days and a major one every six months or so.

In 1886, Edison moved his main laboratory to West Orange, New Jersey, to increase the scale of his research and manufacturing capacity. West Orange was ten times bigger than Menlo Park. Edison's biographer, Josephson, describes the reason behind the move:

> I will have the best equipment & largest Laboratory extant, and the facilities superior to any other for rapid & cheap development of an invention & working it up into Commercial shape with models patterns & special machines ... Inventions that formerly took months & cost large sums can now be done two or three days with very small expense, as I shall carry a stack of almost every conceivable material.

Edison's factory made the parts necessary for research, and research developed and made the machines for large-scale production in the factory. During the development of the phonograph over forty years, the cylinders developed by research were first made of tinfoil, followed by a wax compound, followed by plastic. The eventual primary use of the phonograph was not the one originally imagined. As this technological and market learning occurred, the capacity to quickly scale up production of new configurations helped Edison gain substantial market share. Edison's New York factories at one stage employed over 2,000 people, one of the largest industrial concerns of the time. In contrast to the high-performance workplace laboratories, these mass-production factories operated with extensive division of labour, and the repetitive, unskilled work led to many industrial disputes.

The scale of activities in West Orange inevitably led to greater departmentalization and administration, which took up more of Edison's time. Although it was highly productive, it never matched the extraordinary output of the Menlo Park period. Edison is quoted as saying: 'From his neck down a man is worth a couple of dollars a day, from his neck up he is worth anything that his brain can produce.' He excoriated 'pinheads' and 'lunkheads', and said: 'A man who doesn't make up his mind to cultivate the habit of thinking misses the greatest pleasure in life.' He hired graduates, but generally preferred generalists over specialists, and this is argued by some to have limited the future development of his research organization. His recruitment methods were idiosyncratic. In the early years, he would point applicants at a pile of junk and tell them to put it together and tell him when they had done so. The junk was a dynamo and those who succeeded in assembling it passed the test for employment. In later years, he compiled lengthy general knowledge questionnaires that prospective inspectors needed to pass before being promoted.

Edison's style was to provide staff with a general outline of what he wanted, and then leave them to decide the best ways of achieving the objectives. He is reputed to have said: 'Hell, there are no rules here—we're trying to accomplish something.' One of Edison's staff said: 'Nothing here is private. Everyone is at liberty to see all he can, and the boss will tell him all the rest.'

He 'managed by walking about', advising and encouraging the teams. Edison worked about eighteen hours a day, and the exercise he received walking from one laboratory table to another gave him 'more benefit and entertainment...than some of my friends and competitors get from playing games like golf'. Edison's biographer, Baldwin, has him 'compelled' to 'wander democratically and visibly up and down the aisles, ubiquitous, endlessly snooping, his sleeves rolled up and his unattended cigar ash dropping onto the shoulders of welders and diecutters'.

Staff worked exceptionally long hours. Tesla complained that in his first two weeks he only managed forty-eight hours' sleep. Legend has Edison working for five straight days and nights, but it was probably three, and it was known that the best time to contact him in the factory was after midnight. Another of his biographers, Miller, suggests: 'The capital crime of the Edison laboratory was to go to sleep. This was a source of disgrace, unless the boss could be caught napping, and then they all followed in line.' Various methods were used to dissuade the somnolent, including the 'corpse reviver', a terrifying noise released beside the ear, and the 'resurrector of the dead', which apparently involved setting sleepers alight with a small exploding substance.

It could be dangerous to work for Edison. His chief assistant, Clarence Dally, lost an arm and most of a hand during experiments with fluoroscopy during which Edison nearly lost his own eyesight. The local press reported Edison generously saying that, although Dally was unable to do any work, he would keep him on the payroll.

Josephson revealingly records the reflections of two of Edison's staff. The first, a young job applicant, was told, 'Everyone applying for a job wants to know two things: how much we pay and how long we work. Well, we don't pay anything and we work all the time.' The applicant took the job. The second, a man reflecting on working for Edison for fifty years, told of his sacrifices resulting from long hours at work, including not seeing his children grow up. When asked why he did it, he responded: 'Because Edison made your work interesting. He made me feel I was making something for him. I wasn't just a workman.'

Despite these practices, which seem draconian today, Edison encouraged a creative and productive workforce. Key employees were paid bonuses from profits on inventions, although this incentive did not extend to Nikola Tesla. He socialized with staff with snacks, cigars, jokes, tales, dancing, and singing (see Figure 7). He organized a popular midnight lunch. There was an electric toy railroad to play with and a pet bear. According to management academic Andrew Hargadon:

> The muckers [engineers] would work for days straight in pursuit of a solution, then punctuate their work with late-night breaks of pie, tobacco and bawdy songs around their giant organ that dominated one end of the laboratory.

7. Edison encouraged play as well as hard work. Here workers attend a 'sing' session.

One of Edison's assistants, quoted in Millard, said that there was 'a little community of kindred spirits, all in young manhood, enthusiastic about their work, expectant of great results', for whom work and play were indistinguishable.

Tesla complained of Edison's concentration on instinct and intuition over theory and calculation, and practices at the laboratory did occasionally seem haphazard. When searching for the best material for the light bulb filament, he experimented with unlikely materials ranging from horsehair to cork to the beards of his workers. When the breakthrough came in the carbon filament incandescent lamp, Edison's staff did not realize the extent of their discovery for several months after the event.

Nonetheless, there was focus and discipline. Edison claimed never to have perfected an invention he did not think about in terms of the service it provided, saying he found out what the world needed, then proceeded to invent. Projects had to have a practical commercial application. Famed for his 'guesswork', Edison insisted laboratory assistants kept detailed records of their experiments in over 1,000 notebooks, although this also helped with patent registration and dispute. Experimentation was extensive: 6,000 distinct species of plants, mainly bamboos, were used for carbonized filaments; and 50,000 separate experiments were undertaken in developing Edison's nickel-iron battery. One of Edison's assistants, working closely with his employer, recorded 15,000 experiments into a particular problem. West Orange possessed an extensive library of some 10,000 volumes, and Edison constantly read about biology, astronomy, mechanics, metaphysics, music, physics, and political economy. Although criticized for his disparagement of formal education, he employed two eminent mathematicians, one of whom went on to be a professor at Harvard and MIT. One of his key chemists was known as 'Basic Lawson' because of his adherence to basic scientific principles. Edison met and admired Pasteur and the German physicist and physician Helmholtz, and Einstein gave a seminar at the company. Somewhat incongruously, George Bernard Shaw worked for Edison for a time in London.

Artefacts and drawings were important sources of creativity and communications. Edison is quoted as saying: 'Inspiration can be found in a pile of junk. Sometimes, you can put it together with a good imagination and invent something.' In 1887, his laboratory was reputed to have contained 8,000 kinds of chemical; every kind of screw, cord, wire, and needle; animals from camels to minx; feathers from peacocks and ostriches; hooves, horns, shells, and shark's teeth. Edison found it easier to think in pictures rather than words. When he was contracted by the Western Union Telegraph Company in 1877 to improve upon the telephone invented by Alexander Graham Bell, he produced over 500 sketches leading to his improved design.

As well as his internal efforts, Edison assiduously cultivated his business and research networks. He was a broker of technology, transferring research between industries. As well as his own experiments, he undertook contract research for the telegraph, electric light, railroad, and mining industries. As Hargadon puts it:

> Edison quietly blurred the line between the experiments he did for others and those he did for himself. Who was to know if a result from contract research was applied to another project or if experimental equipment built for one customer was used in work for another.

His ability to continually innovate, according to Hargadon, lay in the way he knew how to exploit the networked landscape of the time.

Edison's approach was one of trial and error, hard work, and persistence; being methodical, rigorous, and purposeful; and using prepared minds and careful monitoring. He believed innovation arose not from individual genius but collaboration, and this capacity to work together and across boundaries resulted from a supportive culture, environment, and social and industrial relations.

Edison worked on the cusp of the transition between the era of the great individual inventor and systematic, corporate organization of innovation. He created a form of organization for the emerging, modern technological society that was rapidly emulated by large corporations such as Bell and General Electric. In an article in the *New York Times* on 24 June 1928, it was estimated that Edison's inventions had built industries valued at $15 billion ($209 billion at 2017 prices). His fame was universal. President Hoover called Edison a 'benefactor of all mankind', and when he died asked people to switch off their lights for a 'minute of darkness' in his memory (Figure 8). His obituary in the *New York Times* on

In the cartoon, text on the document reads: "IN RESPONSE TO THE UNIVERSAL DESIRE TO PAY PERSONAL RESPECT TO MR. EDISON I RECOMMEND THAT ALL INDIVIDUALS SHOULD EXTINGUISH THEIR LIGHTS FOR ONE MINUTE ON WEDNESDAY EVENING OCTOBER 21 AT 10 O'CLOCK EASTERN TIME. HERBERT HOOVER"

Below the cartoon: " Put out the lights, and then—put out the light!'"

8. 'Put out the light' cartoon, by Clifford K. Berryman, 1931. President Hoover called for electric lights to be switched off for a 'minute of darkness' in memory of Edison's achievements.

18 October 1931 began: 'Thomas Alva Edison made the world a better place in which to live and brought comparative luxury into the life of the workingman.' An innovator can make no greater contribution.

Workplaces

As Edison so clearly showed, innovation is more likely to occur in organizations that are forward-looking, accepting of risk, and tolerant of diversity and failure. A workplace that is playful and fun and where conversation and laughter is common is more likely to be innovative than one that is highly formal, bureaucratized, and impersonal. Where expression of opinion is welcomed, ideas are not only generated more regularly, but are also implemented faster. Opposition is voiced when it has a chance to be productive rather than in the subsequent subversion of decisions.

IDEO is a company with a highly innovative workplace that emulates some of the lessons from Edison. It is a successful provider of design and innovation services, employing over 550 people in offices around the world. It has built a reputation for helping other firms to innovate in their products and services by applying creative techniques learnt in the design studio and design school environments. The company combines 'human factors' and aesthetic design with product engineering knowledge to produce thousands of products for firms from Apple to Nike to Prada, and it designed the whale that starred in the film *Free Willy*. IDEO has been described by *Fast Company* magazine as the 'world's most celebrated design firm'; by the *Wall Street Journal* as 'imagination's playground'; and *Fortune* described its visit to IDEO as 'a Day at Innovation U'.

For the company to deal with many diverse projects, it recruits a wide range of talent, and also enjoys special links with the Stanford University Institute of Design. It employs graduates from psychology, anthropology, and biomechanics as well as design engineering.

The leaders of IDEO have a very high profile in the international design community. They claim to have an innovative culture—'low on hierarchy, big on communications and requiring a minimum of ego'—that uses:

> a collaborative methodology that simultaneously examines user desirability, technical feasibility, and business viability, and employs a range of techniques to visualize, evaluate, and refine opportunities for design and development, such as observation, brainstorming, rapid prototyping, and implementation.

IDEO sells its design methodologies to other companies in the form of courses and training materials. It possesses a large repository—a 'toy box'—of devices and designs from a wide range of products which staff play with when seeking solutions to new problems. It is particularly skilled at playing with creative ideas developed for one industry or project to explore their innovative application in others. Playfulness in this environment enables the cross-fertilization and the serendipitous connection and combination of unrelated ideas.

Structures

Edison pioneered a way of organizing, but organizations have wide choices in how they structure opportunities for innovation. Some choose to be highly formal and bureaucratized, others to be informal and unrestricted. Some try to do both, with parts of the organization encouraged to behave very differently from others.

In one of the earliest studies of innovation in organizations, Burns and Stalker in 1961 distinguished between mechanistic and organic forms of organizing. They argued the former is appropriate to stable, predictable conditions and the latter to changing conditions and unpredictable situations. The basic general principle still applies; the way things are organized should be appropriate to the particular circumstances and aims of innovation. When technologies and markets are rapidly evolving and their future uncertain, the need—as in the case of Menlo Park—is to encourage experimentation and creativity without constraining them with bureaucracy. When some of these uncertainties diminish, a more planned approach is needed to the development of projects, with highly prescribed budgets and operations geared up to deliver the innovation. Furthermore, the form of organization used often changes over time as different issues of innovation emerge. As the process of developing the innovation progresses, the supportive organizational structures move from being 'loose' to 'tight'.

R&D. R&D can be structured in very diverse ways. Many leading firms in the past relied exclusively on large corporate laboratories to undertake their research: their own large-scale Menlo Park. The archetype of this form of 'centralized' R&D was Bell Labs, which employed 25,000 at its peak and has been awarded 30,000 patents. It has received six Nobel Prizes for Physics and, among other things, discovered the transistor, digital switching, communication satellites, cellular mobile radio, sound motion pictures, and stereo recording. One of its basic science discoveries led to the development of radio astronomy. Founded in 1925, and based in New Jersey, it was the research group for AT&T before that company was acquired by Alcatel-Lucent. Renowned for its past strength in basic research, like many corporate laboratories, it has progressively moved towards more applied research.

The criticism of this way of organizing from a business perspective is the way research tends to be too divorced from the needs of customers and is generally too long-term in orientation. In contrast, rather than having a central laboratory, other firms 'decentralize' their R&D organization structures, with laboratories located close to particular businesses or customers. The problem with this form of structure is research tends to focus on short-term issues and miss opportunities for more radical or disruptive innovations. To try to gain the benefits of both forms, some firms combine a central laboratory with numbers of decentralized R&D laboratories, but this is a choice open only to a handful of the wealthiest.

Companies such as Intel and Rolls Royce have extensive connections with universities, and often co-locate near centres of research excellence. The challenge for this 'networked' form of R&D organization is that to be receptive to knowledge from external research, organizations need to have the internal capacity to absorb it. They need skills to understand, interpret, and utilize externally sourced knowledge, and often require their own in-depth expertise to attract high-quality research partners.

The organizational challenge in R&D is finding the balance between longer-term research that provides new options and insights into potentially disruptive technologies and research that deals with short-term or immediate well-defined problems. Firms often appear dissatisfied with whatever R&D structures they have. With centralized structures they feel customer needs are relegated in importance, and decentralized structures might miss potentially valuable innovations. When both forms are used, continual tensions exist over relative funding levels and ownership of projects. The problems of networked R&D lie in managing and combining inputs from multiple parties and disputes over ownership of intellectual property rights.

One strategy firms pursue to improve returns to internal R&D and access external collaborators for innovation has in recent years been described by Henry Chesbrough as 'open innovation'. The household products company, Procter and Gamble, is an example of an open innovator. It is a science-based company with a strong internal commitment to research. Its strategy is described as 'Connect and Develop', and rather than being 90 per cent reliant on its own research investments as in the past it aims to source half its innovations from outside the company. The way it combines its own internal research with external connections is indicative of a strategy that attempts to benefit from complementary ways of organizing innovation in the same company.

The rapid growth of research capacity in China and India in recent years has the potential to change the ways many multi-national companies organize their R&D. Firms create laboratories overseas to adopt their products and services to local markets, take advantage of particular local research expertise, and to create international networks of research collaboration. Many US and European firms have established substantial R&D organizations in India and China, especially in information and communications technology. The strategy these firms use can change over time. Ericsson, the Swedish telecommunications company, for example, began to invest in R&D in China in the 1980s because it helped them to win government contracts and was evidence of goodwill and commitment. R&D expenditure was ramped up in the early 1990s to take advantage of cheap research staff and to help adapt Ericsson products to the rapidly growing local market. Recognizing the quality and potential of Chinese researchers, both in the company and in local universities, in the late 1990s Ericsson began to locate R&D for its world markets in China. By the early 2000s, some of its R&D groups around the world were closed down and moved to China, and Ericsson's Chinese research groups became core components of the company's global R&D efforts.

New developments. R&D is one of the ways organizations create options for the future. The ways they organize their new product and service development are crucial for how successful they will be at realizing the future options they have. While R&D is generally the organizational space for scientists and technical specialists, new product and service development usually encompasses a broader range of people, including from design, marketing, and operations. These specialists help address the questions of why and how things are bought, and whether and at what cost they can be made and delivered.

There are many tools and techniques—such as 'stage-gate' systems, which operate a number of stop/go decision points in the development process—available to help plan new products and services. These are designed to help decide between competing projects and to ensure those that progress are appropriately resourced. These tools have limitations: they might be very helpful in managing the process of developing products, but they do not tell you whether they are the right products in the first place. They can also become very procedural and kill initiative.

To overcome the rigidities of bureaucracy, some organizations sanction 'bootlegging', or allowing staff to spend time working on their own projects. By giving people time—which can extend to one or two days a week—outside their formal job commitments, highly innovative companies such as Google and 3M encourage personal motivation to innovate and new ideas to emerge and blossom.

Google and its holding company Alphabet use a variety of arrangements and techniques to encourage innovation, and these continually evolve as new opportunities emerge and dead ends are reached. Google has formal structures for investing in start-ups, launching 'moonshot' technologies such as driverless cars, and working on machine intelligence. Alphabet has purchased DeepMind, a leading AI company, and operates laboratories that research urban innovation, ageing, and health science. Google's organizational methods and incentives include their cafés designed to encourage interactions, concentrated periods of one or five days of focused attention on particular problems, processes for formally encouraging ideas and orchestrating their progress, and informal methods for ensuring ideas are stimulated and made known to senior staff. All these methods are designed to attract talented staff and use their skills to full effect.

Another method used to circumvent organizational constraints to innovation is the so-called 'skunkworks'. Used initially by the Lockheed Corporation to quickly and secretly develop aircraft during the Cold War, the term is used to describe a small, tightly bound group working on a special project with considerable operational discretion within a larger organization.

Operations and production. The ways new products and services are made and delivered have been the focus of considerable innovation themselves. Production, for example, is automated, and operations—the processes for turning inputs into outputs—have seen major innovations in the ways work is organized. Innovation in production and operations has helped create mass markets for affordable, high-quality products and services, such as automobiles, consumer goods and electronics, supermarkets, and hotel chains.

One of the key principles in organizing operations and production is Adam Smith's analysis of the division of labour: how specialization in tasks improves productivity. Henry Ford used the principles of specialization and automation in developing his assembly line producing automobiles for the emerging mass market in the early 20th century. Ford's objective was tighter managerial control over production processes than previous craft forms of production allowed. His solution was the development of the mass-production line with high volumes of standardized products made from interchangeable parts, employing unskilled or semiskilled workers. Management and design were the responsibility of narrowly skilled professionals. The control of work by craftsmen was replaced by management, and the pace of work was dictated by the need to maximize use of equipment. Because machinery was so expensive, firms could not afford to allow the assembly line to grind to a halt. Buffers of extra supplies of materials and labour were added to the system to assure smooth production. Standard designs were kept in production for as long as possible because changing machinery was expensive, resulting in consumers benefiting from lower costs but at the expense of variety and choice.

Ford's friend Edison had already experienced the problem that unskilled, repetitive work caused with the industrial disputes he faced. General Motors showed Ford the limitations of his marketing approach and the benefits of producing varieties of vehicles. Alfred Sloan's approach at General Motors aimed to produce 'a car for every purse and purpose'. But the real innovation allowing both efficiency in production, wide customer choice, and better use of skills came from Japan.

After the Second World War, Toyota recognized that to realize its ambition of becoming an international car-maker it needed to harness the efficiency of American mass-production techniques and craft quality of Japanese work practices. At that time local Japanese automobile markets were small and demanded a wide variety of vehicles, production techniques were primitive in comparison to those in the USA, and investment capital was scarce. Unionized Japanese factory workers insisted they retain their skills and were unwilling to be treated as variable costs, like the interchangeable parts in Ford's and Edison's factories. Toyota understood the dangers of repetitive and boring tasks resulting in fatigue or injury to workers, with diminishing returns to efficiency.

In 1950, Toyota's President, Eiji Toyoda, spent three months at Ford's Rouge factory in the USA. He was amazed at the total output of the plant, which in one year produced over 2.5 times the number of cars made by Toyota in the previous thirteen years. But while total output was impressive, Toyoda thought the system was wasteful in effort, materials, and time. Toyota could not afford to produce cars with such narrowly skilled professionals and unskilled workers tending expensive, single purpose machines with their buffers of extra stocks and re-work areas. Toyoda's objectives were to simplify their production system combining some advantages of skilled craft working with those of mass production, but avoiding the high costs of craft and rigidities of factory systems. The result was the evolution of Toyota's lean production system employing teams of multi-skilled workers at all levels of the organization and highly flexible, automated machines producing large volumes of highly varied products. Rather than having buffer stocks of inventory, thereby wasting resources, Toyota's system delivers components just-in-time to be used.

Teams of Toyota workers are given time to suggest improvements to production processes in 'quality circles': Toyota has several thousand quality circles that complete tens of thousands of small improvement projects every year. Quality circles are linked to efforts for continuous improvement (*kaizen*) in collaboration with industrial engineers. Emphasis on problem-solving is an important part of everyone's job and on-the-job training, collective education, and self-development are all encouraged.

The success of lean production improved the whole system of designing and making cars and it made Toyota the car producer against which other manufacturing firms compared themselves. The combination of technical and organizational innovation in Toyota's production system has produced both economies of scale and scope: volume and variety.

The search for innovations that help combine economies of scale through standardization with economies of scope to satisfy diverse consumer choices is a continuing challenge. The ultimate objective in many cases is producing economically for markets of one.

Services organizations similarly look for innovation in their operations. Airlines use Internet-based booking and ticketing systems, and apps manage boarding passes and frequent flier rewards. Supermarkets mine the data about purchasing behaviour from loyalty cards to target sales promotions and make sure goods available in stores are tailored to fit the profiles of local customers.

As the world's largest ecommerce company by revenue, Amazon is an innovation leader in operations and logistics. The company began selling books in 1994 and now has over 300 million active customers buying millions of different products. It exceeded $100 billion sales in 2015 and continues to grow rapidly. Its founder, Jeff Bezos, has been ranked as the world's wealthiest person. Being reliable and quick at dispatching goods is key to Amazon's success. The company can dispatch goods within fifteen minutes of receiving an order.

The company is highly innovative and offers services such as Amazon Prime that provides free delivery for a fee and access to streamed films. It sells products such as e-book Kindle and Alexa, the voice-controlled personal assistant. Amazon is the world's largest producer of cloud infrastructure services, which are considered so secure they are used by the CIA. It is exploring deliveries by drones, shops where your Amazon account is automatically debited as you walk out of the store without the need to pay at a till, the delivery of food, and, ironically given it was the first to sell books online, it is opening book shops. It is exploring opportunities in healthcare and the delivery of pharmaceuticals and offers a cloud-based business analytics service that builds visualizations to provide business insights.

Technology is critical for the operation of its warehouses, some of which are as big as twenty-eight football pitches. The use of space in warehouses is optimized for maximum density and height of storage, assisted by the use of 45,000 robots. Amazon acquired the company that made these robots and according to some accounts is adding 15,000 robots each year. Post-acquisition the company stopped selling robots to Amazon's competitors, highlighting the commercial advantage they deliver. In the past, goods moved around warehouses using conveyor systems and forklifts, and workers walked between stacks to locate goods for dispatch. Nowadays, all products have barcodes or radio frequency identification devices and are entered into a database that locates the nearest robot once an order has been received. Using sensors to navigate and avoid collisions robots move to the identified stack with the product and then lift and carry it to an operator for packing.

Amazon has been an extraordinary business success story, and this continues with its move into Web services, where its cloud computing and hosting revenue quickly produced multi-billion-dollar annual revenue. It has faced criticism over its employment practices and the impact it has had on small high street stores. It has, however, fundamentally changed the way many companies sell and people shop, and this has depended on innovation in its operational processes.

Alibaba is the world's largest online retailer by volume of sales, and its success is also based on a business model using digital technologies. Jack Ma, its founder and executive chairman, explains the company's approach to the use of technology is that whereas IT aims to control, data technology aims to share. Its intention is therefore to empower rather than control a trading ecosystem. Alibaba collects and refines data on where and what shoppers are buying and looking at online; it then feeds that data back to merchants to better target their sales; and uses predictive data to let merchants know what they should stock and make. This encourages more use of Alibaba's ecommerce platforms. The capacity of Alibaba to deal with extraordinary amounts of data is seen in a peak day of sales in China where its site was processing 175,000 transactions every second. The company has also bought physical stores and is collating data on shopping habits in both the online and physical world, providing insights for even more efficient inventory management.

Networks and communities. Edison's development of the electrical lighting industry was an example of innovation in a technical system brought about within a network of innovators. Most innovation involves the participation of numbers of collaborating organizations and from the perspective of the individual organization this brings benefits and difficulties. The benefits lie in being able to access knowledge, skills, and other resources it does not itself possess. The difficulties lie in the absence of organizational sanction in getting others to do as you wish.

The key to effective networking is building partnerships with high degrees of trust. Trust is needed in the technical competence of collaborators, their ability to deliver what is expected of them, and their overall integrity in protecting proprietary knowledge and being prepared to admit when things go wrong. Collaborations usually start as a result of personal connections. These can break down as people move to other jobs or organizations. Effective trust between partners therefore involves the extension of inter-personal trust to inter-organization trust, with the value of the collaboration becoming institutionally engrained: legally, administratively, and culturally.

In some fields, such as open-source software, the community of users is the innovator. Here it is users of the product or service that provide new content and improvements. Despite the rhetoric of unconstrained engagement in many of these communities, a degree of organization is required. Wikipedia, for example, recognizes the efforts of its contributors to its online encyclopedia by creating a hierarchy, with significant community status accorded to wikipedians whose contributions have reached high levels of quality and quantity.

Organizations are becoming more adept at using social networking sites, wikis, and blogs in their innovation activities. They are using social networking analyses by, for example, surveys or tracking email correspondence, to understand the key personal and organizational nodes in the organization and help improve decision-making.

Projects. A large part of modern economies is comprised of large, complex infrastructural projects, such as telecommunications networks, energy production and distribution, and transportation systems of airports, railways, and motorways. These projects, which commonly cost billions of dollars, entail the coordination of large numbers of firms that assemble to contribute their various skills and resources during different stages of the project's progression. They are notorious for cost over-runs and delays. The Channel Tunnel, between England and France, for example, was 80 per cent over budget. Innovation provides the means by which these projects can deliver on expectations.

London Heathrow Airport's Terminal 5 (T5) was a large and highly complex project, with a budget of £4.3 billion and involving over 20,000 contracting organizations. Overseen by the British Airport Authority (BAA), the project client, airport owner, and operator, it entailed the construction of major buildings, a transit system, and road, rail, and subway links, alongside the world's busiest airport working at overcapacity. T5 is the size of London's Hyde Park, and has an annual capacity of thirty million passengers. Although often remembered for the disastrous first few days of operation where British Airways misplaced 20,000 bags and cancelled 500 flights, the design and construction of the project itself was a success and was delivered to budget and on time. This success resulted from an innovative approach to managing large, complex projects.

BAA took care to learn lessons from previous projects, made sure any technologies used were already proven elsewhere, and trialled new approaches on smaller projects before applying them to T5. Use was made of digital simulation, modelling, and visualization technologies to help integrate designs and construction. Underpinning the success of the T5 project was a contract between the client, BAA, and its major suppliers that differed considerably from industry norms—which were commonly adversarial—and encouraged collaboration, trust, and supplier responsibility. The risk in the project was assumed by BAA, work was conducted in integrated project teams with first-tier suppliers, and incentives were designed to reward high performing teams. Although the processes and procedures to be followed were highly specified, the project was formulated in a way that allowed managers to confront the unforeseen problems that inevitably arise in complex projects flexibly and on the basis of their previous experience.

The lessons from T5 are that success in large, complex projects involves standardized, repetitive, and carefully prepared routines, processes, and technologies and the capacity to be innovative to be able to deal with unexpected events and problems. Organizing projects involves a judicious balance between performing routines and promoting innovation. These lessons have been applied in subsequent projects such as Crossrail, the major new railway traversing London.

Creative people and teams. As Edison showed in Menlo Park, innovation involves a team effort, bringing together different ideas and expertise. The construction of teams involves decisions about the most appropriate balance of skills given the problems faced. It also entails deciding on the comparative value of organizational memory—keeping people in teams together—and refreshment—bringing in new skills. Teams that work together for long periods tend to become introspective and become immune to innovative ideas from outside. Teams that are newly constructed or contain many new members have to learn to work together effectively and develop a modus operandi. Harmony in teams holds many virtues, but sometimes it is important for innovation to have disruptive elements—the grit in the oyster—asking difficult questions and shaking things up.

The structure of teams has to reflect their objectives. Those devoted to more radical innovation need more creativity and flexibility in objectives, with the freedom to respond to emerging and potentially unforeseen opportunities. They often need heavyweight support from higher levels in the organization as their objectives do not quickly add to the bottom line and as a result they are vulnerable to criticism and cost-saving exercises. A balance has to be found between incentives for individuals and teams. The factors that encourage innovation team effectiveness are often subjective—to do with professional satisfaction and recognition. Those that inhibit performance are more instrumental—to do with project objectives and resource limitations. As Edison found out, employees will work extraordinarily hard when given the incentive of interesting, rewarding, and appreciated jobs.

Creativity is not only important to design companies such as IDEO. Innovation in all organizations relies on creative people and teams to produce new ideas, and creativity is an issue that infuses the whole of the world of work. By its stimulation of innovation, many contemporary organizations see the encouragement of creativity as core to their development and competitiveness. Creativity provides a means of making work more attractive, improving the engagement and commitment of existing staff, and a winning strategy in the 'war for talent' among highly skilled and mobile employees.

Creativity has an individual and a group component. Psychologists tell us about the characteristics of creative people, and how imaginative ideas emerge from individuals with the capacity to think differently and see connections and possibilities. Creative individuals are said to have a tolerance for ambiguity, contradiction, and complexity. Cognitive scientists, such as Margaret Boden, argue that creativity is something that can be learnt by everyone and is based in ordinary abilities that we all share, and in practised expertise to which we can all aspire.

Organizations expend large amounts of time and resources on creativity training, and constructing incentives and rewards for individual creativity. They are also concerned with the promotion of creativity in groups and with formulating the most conducive team structures and organizational processes and practices. Groups bring together the disparate perspectives and knowledge that are valuable for creativity and are essential for the new combinations in innovation.

Creative ideas become useful innovations when they are successfully applied. Creativity in itself may be inspiring, stimulating, and beautiful, but it has no value economically until it is manifested as an innovation. It takes different forms in incremental and radical innovations. Incremental innovations usually involve a form of creativity that is more structured, managed, and deliberate. Radical innovation requires creativity that may be unbounded by existing practices and ways of doing things.

People

Leaders. Innovation rarely occurs in organizations without the commitment and visible support of their leaders, although those leaders may have little idea of the specific nature of new developments. One of the key aspects of leadership is encouragement for the creation of new ideas and their implementation. Leaders find resources for support and offer protection from the opponents of innovation, and give employees permission to get excited about new ideas. When new ideas threaten the status quo, established interests will inevitably oppose them. As Machiavelli says in *The Prince*:

> There is nothing more difficult to plan, nor more dangerous to manage than the creation of a new order of things...Whenever his enemies have the ability to attack the innovator they do so with the passion of partisans, while the others defend him sluggishly, so that the innovator and his party alike are vulnerable.

One of the lessons of renowned leaders of innovative organizations, such as Edison, is they create a supportive culture where staff are encouraged to try new things and are not discouraged when they fail. In 1948, the chairman of 3M, William McKnight, summarized his approach that characterized the company's strategy for decades to follow...

> As our business grows, it becomes increasingly necessary to delegate responsibility and encourage men and women to exercise their initiative. This requires considerable tolerance. Those men and women to whom we delegate authority and responsibility, if they are good people, are going to want to do their jobs in their own way.
>
> Mistakes will be made...Management that is destructively critical when mistakes are made kills initiative. And it's essential that we have many people with initiative if we are to continue to grow...

A nervous young manager who had led a failed project once offered Henry Ford a resignation letter. Ford's response was that he was not about to let someone go and work for a competitor after learning a valuable lesson with his money.

Managers. As well as supportive leadership at the top of organizations, particular innovations need enthusiastic and powerful managerial 'champions' or sponsors with significant decision-making responsibility. As well as being good at managing teams, coordinating technical/design issues, and implementing processes and decisions, innovation managers also have to be skilful at advocating the virtues of the innovation, lobbying for its support, and creating a vision of what it will do and contribute.

Boundary spanners. One of the most important individual roles in innovation is the boundary spanner, the person capable of communicating and building bridges between and within organizations. In manufacturing firms, this person used to be known as the technological gatekeeper. These people are avaricious acquirers of information—attained through reading and attending conferences and trade shows—and skilled at communicating useful information to the part of the organization that needs it. Organizations sometimes find it difficult to justify appointing boundary spanners. Their remit to travel, go to conferences, and talk to lots of people is sometimes unappreciated by those bound to a desk or workbench. But their role is highly beneficial to innovation.

Everyone. One of 3M's most successful innovations is the Post-It Note. Due recognition has been made of the developers of the technical core of the innovation—its non-sticky glue. And sufficient opprobrium has been dealt to the company's marketing department that claimed no one would buy it. But too little credit has been accorded to the people in the organization who recognized the product's potential and encouraged its development. Following the rejection of the idea of the Post-It from the marketing department, the product's developers sent samples to the secretaries of the company's general managers. The secretaries immediately saw the value of the product and elicited the support of their bosses in getting the idea developed.

Innovation affects everyone in an organization and to a greater or lesser extent is everyone's responsibility. The computerization of many traditional engineering craft skills, such as toolmaking, provided opportunities for de-skilling or re-skilling jobs. Many employers took the de-skilling path—as in the case of numerical control machine tools—but subsequently learned the advantages of re-skilling and giving the shop-floor worker discretion over the tasks they perform. This reflects the capacity of people to change, and productively and creatively respond to innovation, if given the chance. It gives them, in Edison's words, the pleasure of cultivating the capacity to think. The potential of innovation derived from the factory floor has led some to describe them as laboratories and places for experiment.

An important tool used to encourage innovation is the use of reward and recognition programmes. Many organizations have suggestion schemes, and companies such as IBM and Toyota elicit hundreds of thousands of ideas from employees. These can be rewarded financially or by peer recognition. Often the most effective form of recognition is the implementation of the idea by the organization. The capacity of individuals throughout the organization to have innovative ideas and pursue their implementation shows that innovation leadership is not only the responsibility of those with high hierarchical positions.

Innovators in all forms are best supported in organizations whose commitment to human resource development and training attracts, rewards, and retains talented managers and staff unafraid of change, and placates those who are fearful of it. Innovative organizations have the appointment procedures, payment and incentive systems, and career progression paths, to ensure appropriate staffing for innovation. While some people thrive on creating innovation, and need to be encouraged and rewarded, others are better at developing the procedures for its application, requiring different forms of recognition. Others still are temperamentally fearful of innovation, or at least too much change, seeing it as threatening, and can suffer stress and poor performance as a result. A reputation as an innovative organization is very attractive to potential recruits who want to be innovative, and selection mechanisms should vet unsuitable appointments. Employees who find it disturbing need to be supported and guided through the introduction of innovation.

Technology

During the 1960s, the research of Joan Woodward into factory organization in the southeast of England began to explain the relationship between technology and organization. She showed how organization varied according to core underlying technology, whether production took the form of small or large batches, mass production, or continuous flow processes. The view that organization results from the technology used—technological determinism—has been discounted by research showing the extent to which choices can be made, a view to which Joan Woodward subscribed. Nonetheless, technology is highly influential, and there is a relationship between the ways industries are organized and the extent to which they can benefit from innovation through the division of labour. The products and services of industries vary considerably and production and operations techniques vary accordingly.

Innovation technologies. Edison knew the value of high-quality scientific instrumentation, on the one hand, and 'junk', odd pieces of machines, and a vast array of unusual materials, on the other. These machines and artefacts stimulate innovation. Just as Edison's many sketches aided his thinking and improved the communication of his ideas with others, the creation of tangible designs and prototypes focuses efforts and builds connections between people with different skills and perspectives. In many cases, ideas for innovation grow organically and iteratively around emerging and increasingly focused designs.

New technologies move design and connections across boundaries into a digital world where Edison's aim for 'rapid and cheap development of an invention & working it up into Commercial shape' occurs in ways he could not imagine. Digital design information on new products is transferred to the equipment used to make them. The designs are guided by the system as to what is possible to manufacture (see Figure 9).

Developments of massive computing power, software that allows the merger of different data sets, and new visualization technologies used extensively in the computer games industry, have led to a new kind of technology supporting innovation. 'Innovation technology' (IvT) is so called because it helps combine various components of the innovation process. It is being used

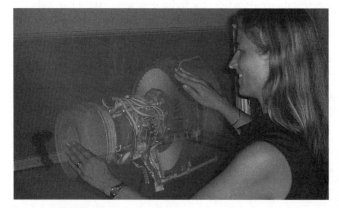

9. Touchlight: engineering and design increasingly use computerized visualization and virtual reality tools.

to improve the speed and efficiency of innovation by pulling together different inputs within and between organizations. IvT includes: virtual reality suites used to help customers design new products and services; simulation and modelling tools used to substantially improve the speed of new designs; data science, building new communities of scientists and researchers, and helping them to manage collaborative projects; AI, machine learning, and sophisticated big data-mining technology used to help improve research, understand customers, and manage suppliers; and virtual and rapid prototyping technology used to improve the speed of innovation. Together, these technologies are being used to make more effective decisions about innovation.

By moving experiments and prototyping into the digital world, IvT allows firms to experiment cheaply and 'fail often and early'. IvT is also very important in the design of large, complex systems, such as utilities, airport infrastructures, communications systems, where it is not usually feasible to test full-scale prototypes.

One of the most important aspects of IvT is how it assists the representation and visualization of knowledge, and its communication across different domains, disciplines, professions, and 'communities of practice'. By way of illustration, compare the design of a new building using traditional methods and IvT. The use of IvT makes complex data, information, perspectives, and preferences from diverse groups visible and comprehensible. Virtual representation assists architects to visualize their eventual designs and help to clarify clients' expectations by giving them a good understanding of what a building will look and feel like before work begins. Clients can 'walk through' their virtual building getting a sense of its layout and 'feel' prior to a brick being laid. IvT informs contractors and builders of specifications and requirements, and allows regulators, such as fire inspectors, to confidently assess whether buildings are likely to meet regulatory requirements. IvT can allow various players in the innovation process, suppliers and users, contractors and subcontractors, systems integrators and component producers, to collaborate more effectively in the delivery of new products and services.

Using IvT can produce some quite dramatic innovations. A great many died in the World Trade Center in 2001 because occupants trying to get down the fire escape stairs were trapped with firefighters going up them. New ways of getting people out of tall buildings in extreme events were considered for the replacement Freedom Tower in New York. Computer simulations and visualizations of the behaviour of buildings and people in emergencies led fire engineers to believe that the safest form of egress was by means of the lift. Changing entrenched views on safety to one where the message is: 'In case of fire, use the lift' requires a great deal of persuasion of building owners and occupants, engineers and architects, firefighters and fire regulators, and insurers. Mutual, shared comprehension of this radical change was assisted by moving detail from complex drawings and data sets to readily understandable computerized images. Fire engineers have used a range of new simulation and visualization technologies to help transform these diverse parties' understanding of safety in tall buildings and encourage the exploration of innovative approaches to rapid evacuation.

Chapter 6
Innovating the future

We began this book with an illustration of innovation at the beginning of the Industrial Revolution. We end it with a speculative glimpse at what the future may hold. The challenges and opportunities for innovation are immense. As well as creating new sources of wealth from ideas, innovation is essential if we are to cope with climate change, provide better water and food, improve health and education, and produce energy sustainably. It will be essential for our continued co-existence on an increasingly crowded planet.

The innovation processes that will be used have become increasingly more complex. They have evolved from the activities of 18th-century entrepreneurs such as Josiah Wedgwood, the formal organization of in-house research in the 19th century begun by Thomas Edison, and the large corporate R&D departments of the mid- and late 20th century in which Stephanie Kwolek worked. Nowadays, innovation involves multiple contributors in distributed networks supported by new technologies, with heightened intensification of all five models of innovation described in Chapter 2: the push from science, appreciating and responding to demand, better coupling of all contributors to innovation within organizations and collaboration with those outside, and enhanced strategic integration and networking using digital technologies.

The keys to future innovativeness will lie in an organization's ability to foster creativity and to make decisions and choices on the basis of being well-prepared, informed and connected. It will lie with the ability of entrepreneurial individuals and teams to see opportunities and take risks and build new types of venture. The many sources of ideas—employees, entrepreneurs, R&D, customers, suppliers, and universities—will continually produce opportunities to innovate. The challenge lies in encouraging, selecting, and configuring the best ideas from them. Crucial behaviours include the appetite to explore and experiment, to play with ideas, and the fortitude and resilience to see them turned into useful innovations.

Enterprises

When economies and technologies are rapidly changing and volatile, the value of firms' capacity to accept and implement radical and disruptive ideas increases. This will prove challenging (see Figure 10). In such circumstances, the best strategies are those that are experimental and dynamic and achieve a judicious balance between exploiting existing ideas and exploring new ones. These strategies rely on continual investments in human capital and in research and technology.

Innovation, in the words of Lou Gerstner, previous CEO of IBM, needs to be engrained in the DNA of the organization. The performance of exceptional innovators and teams should be rewarded, but the responsibilities and opportunities for innovation are everyone's.

Continuing investments in R&D and the absorptive capacity it produces remain crucial, as does the ability to trade and broker knowledge within innovation systems. Corporate connections to sources of new ideas require long-term partnerships with universities around the world, deep embedding within innovative cities and regions, and effective management of supportive IvTs.

'This really is an innovative approach, but I'm afraid we can't consider it. It's never been done before.'

10. Some challenges of innovation may always be with us.

The breadth of these systems is extended as traditional distinctions between industries become blurred as knowledge, insights, and skills across sectors are transferred and combined to produce novel offerings. Much value creation in the manufacturing industry, for example, lies in design services. The services sectors and universities are collaborating in innovative ways. Innovation is a condition for success in the creative industries—such as new digital media, entertainment, and publishing—whose content is critical, for example, for innovative product and services companies involved in mobile telephony. The resource industries, such as agriculture and mining, depend on innovation to improve efficiencies and assist product improvements, and whose innovations in water management, for example, have broader applications.

111

The ideas for innovation in business will come from diverse and often unexpected sources in new and unforeseen combinations. New forms of governance of risk will be needed to oversee greater ethical and responsible decision-making and improve risk management of complex innovations.

Small and medium-sized organizations may increasingly be the progenitors of breakthrough technologies, using their advantages of speed, flexibility, and focus over larger organizations. Compared to large, publicly traded companies, small firms can take and bear unusual risks. And not being so constrained by the organizational rigidities of large firms, they can more easily develop and trial novel business models and processes. Small and medium-sized organizations will combine their behavioural advantages with the greater resources found in large firms in new forms of innovation network and collaborative partnership. Larger organizations will continually experiment with attempts to emulate the entrepreneurial environments of smaller units. There is a trend for many larger organizations to become flatter in structure, in other words with fewer levels in their hierarchy. This is assisted by technology that provides more easily accessible information and better communication between organizational units and levels, allowing decisions to be made further down the organization than in the past.

Another future organizational trend is disintermediation, meaning that providers and users of services will be directly connected through technology. So, new websites for international currency exchange remove the need for this service in banks or currency exchange kiosks in airports, with their extortionate spreads of rates. Charitable giving can be directed to individual recipients rather than going through a charitable organization.

As Edison knew, innovation has to be organized in ways appropriate to its objectives. The benefits from the unbounded search for ideas, where chance and serendipity can produce so many rewards, have to be balanced with organizational focus and direction. There are far more opportunities than can be afforded, and choices need to be made that shape and direct the skills organizations use and resources they invest. Skills in the strategic management of innovation which help them make those choices will become among the most prized by business.

Institutions

Governments. As well as pursuing the innovation policies discussed in Chapter 4, encouraging the stock and flow of innovation, national governments need high levels of inter-governmental coordination: internationally, regionally, and locally.

The use of innovation to deal with many contemporary problems requires more resources and skills than can be mustered by individual nations. Some challenges, such as controlling greenhouse emissions, managing nuclear energy, pharmaceutical regulations, managing cyber-security and terrorism, simply cannot have national solutions and have to be addressed in international forums. Balancing national self-interest with the need for international approaches will pose an increasing innovation policy challenge. Furthermore, as social well-being and economic prosperity become driven ever more by the productive use of creativity and knowledge, there are profound implications for relationships and disparities between nations. Existing inequalities may become accentuated as the technologically, institutionally, and organizationally rich nations pull further away from those not so advantaged. Inter-governmental institutions have to monitor and consider policies to address any such problems.

Many important decisions about innovation are made not at the national level but by increasingly powerful municipal authorities and regional governments that vigorously compete with each other domestically and internationally to attract investment and talent. Expertise in domestic inter-governmental coordination and collaboration is also essential to effective innovation policy. The privatization in many countries of previously publicly held assets in energy, transportation, and telecommunications has removed a direct lever governments once possessed to improve innovation. Instead, new regulatory authorities have been established and their roles in supporting innovation in the private sector have to be explored and extended. One of the greatest challenges for governments is to stay abreast of technological changes and their implications, and this is incredibly hard for emerging technologies such as AI. But full appreciation of any adverse consequences of technology is needed for its effective regulation. Regulations are often decried by the private sector as constraining initiative, and this has to be a concern for government, but they can also stimulate innovation, for example, in demanding zero emissions from cars or energy loads from renewable energy sources. Government's role is complicated by the ways boundaries of the public and private realms have become blurred with the creation of public/private partnerships. There are mutual benefits to be achieved from this form of organization, including access to resources for investment in innovation that may otherwise be unavailable. But the ownership and control of innovation assets and knowledge may be shaped by different incentives that may result in tensions between private gain and public good. Government innovation policy has to be formulated on the basis of deep engagement with business, and understanding of the strengths and shortcomings of the contributions it can make. Much innovation can be unleashed when governments make the data they collect available, suitably anonymized, to entrepreneurs and organizations to produce imaginative new products and services.

There are huge future opportunities for innovation in government services. These include, for example, health and well-being technologies, using wearable monitoring devices and mobile phones to assist medical diagnoses at home and provide monitoring of elderly patients preventing them having to be in hospital. Telemedicine is used in Australia to deliver health services to remote communities. In India, mobile equipment is driven to impoverished villages where diagnoses are made via electronic links to urban hospitals, providing a level of healthcare to which the rural poor previously had no access. AI can be used to predict when crises might emerge, for example by foreseeing famine through predicting crop failure and allowing the pre-emptive delivery of aid.

As part of their contributions to innovation, governments have opportunities to use the new technologies that offer the means for more inclusiveness and participation in decision-making by citizens in designing and delivering the services we demand. By creating proposed new health centres in virtual technologies prior to their actual construction, for example, inputs can be elicited from health professionals and patients to produce better designs.

One of the most crucial areas of policy-making lies in the processes by which choices are made by governments about where to focus innovation investment to sustain future prosperity. No nation has the resources to innovate in all areas, and trade-offs are needed between competing demands for scarce resources. Governments have to establish sophisticated approaches for making choices about the 'grand challenges' they face, while ensuring that enough is invested in a broad spread of areas to keep options open and allow nations to absorb useful ideas developed elsewhere. Decisions about what and what not to prioritize have to involve extensive discussions with business, social, and environmental groups, and informed public debate in efforts to build consensus about the challenges of the future.

The importance of innovation for government, and the difficulties in building the necessary connections and making good choices, requires broad and deep innovation policy-making skills. These extend comprehension of the importance and nature of innovation throughout the apparatus of government, and help develop a 'whole of government' approach. Greater appreciation of the contributions and difficulties of innovation will help address the very high aversion to risk in the public service. In recognition of its more widespread, distributed, and inclusive nature, public policy requires better forms of measurement of innovation—moving on from the partial and often misleading indicators of R&D expenditure and patenting performance—and new approaches and skills are needed in this area. Tools, such as social network analysis can be used, for example, to measure changing patterns of connectivity. Innovation policy-making has to recognize that innovation is a continuing challenge with no simple 'solutions'. As it evolves new issues arise and policies need to change in response.

Universities. Universities are core hubs of the world's economically fastest growing districts. They research to make discoveries and are increasingly addressing large complex challenges through connecting expertise in multi-disciplinary and international centres. Scientific attention is increasingly closely coupled with pressing social need, and this is seen in their work in the life sciences, in areas such as synthetic biology, bioengineering, and nutrition science, and in addressing severe problems such as microbial resistance. Their educational offerings are extending from priming students to be the leaders and professionals of the future to preparing them for as yet unknown jobs and work. As well as encouraging innovation through their research and teaching, universities invest in knowledge exchange and external flow of ideas. They should welcome the many opportunities collaboration provides for the creation and transfer of new educational and research services, and move beyond a restricted model of 'technology transfer' in the form of formal intellectual property protection, licensing, and start-up companies. Their strategies will need to find multiple ways of engaging with stakeholders in business, government, and the community, and yet still continue to be driven by scholarly values. They educate and employ people able to work in multiple ways in research, business, and government, and build connections between different parts of innovation systems, encouraged by the mobility of broadly skilled graduates and enhanced through the use of digital technologies.

Universities have a continuing role to play in producing the large-scale research tools and instruments for science and engineering to encourage discovery and data exploration, allow people to scout in unknown areas, and to see and measure the things that others cannot. They provide leadership in formulating common standards needed by innovators to help launch new products and services in dynamic industries. They seed the creation of new ventures and industries.

The provision of 'rehearsal space' and collaborative laboratories for playful interactions, deep and sustained conversations, and engagement in ideas generation and testing with business, government, and the community is one of the most important innovation supporting roles for universities. Researchers will continue to work with all the academic rigour and independence of their discipline, but through these conversations many will become comfortable as members of distributed teams exploring inter-disciplinary interfaces, and the social and economic consequences of their work. Highly accomplished at providing the physical and organizational structures and incentives for scholarly recognition and career progression, universities will need to explore better spaces and methods for convening and rewarding such engagement.

Innovating innovation

As in the time of Wedgwood, innovation will result from the combinations of ideas, but these ideas are ever more widespread and distributed around the globe, and their integration can increasingly be assisted by the use of technology. As Wedgwood understood so well, innovation combines 'supply side' considerations—that is, sources of innovation such as research and technological developments—and deep appreciation of market demand. Smart innovators are immersed in understanding the changing patterns and meaning of consumption, and the values and norms that underlie decisions to purchase innovative products and services. These patterns are affected by globalization and are fluid in nature. A generation raised on conspicuous consumption, whatever its real costs, may be despised by another concerned for sustainability. Recognizing the capacities of new technologies to include increasing numbers of contributors to innovation, including communities of users, greater appreciation is needed of their motivations and how their energy and insights can be most effectively used.

Innovation strategy in companies of all sizes and sectors has to move beyond the planned, sequential models of the industrial age, and the forms of corporate R&D laboratories that produced Stephanie Kwolek's discovery. It has to account for opportunities that arise in unexpected places, high levels of uncertainty, and great complexity, where organizational learning through collaboration is the key to survival and growth. The limited financial measurements and accounts used by business in the past—such as return on capital and quarterly reports to shareholders—have to be supplemented with indicators more meaningful to innovation and organizational resilience. What, for example, is the value of the options for the future that organizations possess through doing research? What innovations being explored and developed have the potential to account for major parts of the organization in ten to twenty years? How has an organization's capacity to learn been improved through research investments? What is the value of being a trusted collaborator, and ethical employer and sustainable producer?

Economic thinking benefits from evolutionary approaches that see risk, uncertainty, and failure in innovation as normal and move us away from linear and planned to open, adaptive, and highly connected systems. The value of ideas and learning becomes recognized as the most important driver of economic growth and productivity. The importance of the exploration of new inter-disciplinary combinations between science, arts, engineering, social sciences and humanities, and business is appreciated; and the need for mechanisms and skills for building connections across organizational, professional, and disciplinary boundaries is emphasized. Attention is placed on improving the connections and performance of innovation systems and evolving ecologies of organizations. These ecologies may form unimagined new combinations: anthropology may inform local energy production and distribution; philosophy may influence semiconductor circuit design; the study of music may affect the provision of financial services.

IvTs intensify innovation. The instrumentation of trillions of devices and sensors embedded in the physical world contribute to the unimaginable amounts of data available to be used by the new technologies of design in the virtual world to create and improve the products and services we want and enhance the experiences we desire.

Innovation has to deliver non-damaging or environment-enhancing products and processes. Innovation and sustainable development will need to become two sides of the same coin. Many sustainability challenges—climate change, water resources management, genetically modified agriculture, waste disposal, marine ecosystem protection, and biodiversity loss—are persistent and have no complete solution, lacking a clear set of alternatives and little room for trial and error. They are characterized by contradictory certitudes among protagonists, and strategies for dealing with them involve coping rather than solving and searching for what is feasible not optimal. Lessons from the study of innovation can be applied to deal with these persistent problems, including the facilitation, structure and management of cooperation, and connectedness, the management of risk and assessment of options, and the use of collaboration tools such as social networking technologies. Furthermore, the use of IvTs can help model and simulate the implications of decisions, and their visualization capacities help communications and the informed involvement of diverse parties to assist participative decision-making.

More innovations are being developed that are 'inclusive' in the way they emerge 'bottom up' rather than 'top down', empowering the people who face particular economic or social problems to shape and design their own solutions to them. Smartphones, broadband networks, cloud computing, and blockchain (a tool that securely stores information in a shared record of blocks on information in the cloud) are combining to produce a cheap platform with universal reach available for the first time to billions of the world's poorest people. Apps are emerging, for example, that advise slum dwellers and impoverished rural farmers about market prices, water availability, and medical services.

An example of innovation that can aid inclusiveness is the advent of digital money accessible on mobile phones. It is a technology that moves economic transactions from the physical into the digital world, and can help end the divide between those who can and those who cannot participate in formal economic transactions. With around six billion mobile phones in use, just about everyone in the world, rich and poor alike, now has access to mobile devices. Fewer than two billion of the world's seven billion population, however, have bank accounts, so the vast majority of people are excluded from the banking system. Banks and financial service companies are using mobile technologies to allow digital money transactions for millions for the first time. The availability of micro finance on mobiles overcomes the problems of reach into rural and impoverished areas where there are no bank branches or ATMs, and allows the previously disenfranchised to participate in the broader economy. People can now pay others small amounts using cheap mobiles. Digital money will alleviate the drudgery of dealing with bureaucracy, such as the relentless queuing for paying utility bills, and from having to meet face-to-face to transact, providing the great resource of time.

To use and benefit from these technologies you need a digital identity. Through its Unique Identification project India aims to issue each resident a twelve-digit unique number, which will be stored in a centralized database and will be linked to basic demographic and biometric information. This project has already registered over one billion people, and among other benefits for the poor and underprivileged, it gives them access for the first time to the many services provided by the government and the private sector. While governments have the potential to use these technologies to save money and provide better public governance, society has to deal with issues of data security and privacy, which is challenging enough in the developed world let alone for those engaging with digital technology for the first time.

Automation and the future of work

The relationship between innovation and automation is important for the future of work and for some of society's most pressing concerns, including rising inequality. There are widely differing views on the impact of new forms of automation such as AI and robots. For some they presage a dystopian future of mass unemployment and meaningless, de-skilled jobs; while for others they offer a liberation from menial and dangerous tasks, allowing more creative and meaningful work. The professional services firm PwC estimates that global GDP could be up to 14 per cent higher in 2030 as a result of AI—the equivalent of an additional $15.7 trillion, providing a massive new commercial opportunity.

It is helpful to return to Schumpeter's dictum about innovation being a process of creative destruction. Take the case of autonomous vehicles, or driverless cars. Over 1.2 million people are killed on the road worldwide each year, and the internal combustion engine is a major source of pollution. New technology in driverless, electric cars may significantly reduce this human and environmental carnage, and may create novel sectors and jobs in fields such as software, visualization, and battery production and storage. Yet driving lorries, vans, and taxis is one of the major forms of employment, and the motor industry that makes and maintains petrol-driven vehicles is a crucial component of many national and local economies. The new digital technologies could massively increase unemployment among those who would find it difficult to find alternative work, causing immense social upheaval. At the same time, they could also reconfigure work so that drivers control several vehicles simultaneously and do so remotely, so new kinds of jobs are created. Autonomous vehicles exemplify Schumpeter's view that innovation simultaneously creates and destroys. Social progress from innovation involves accentuating the creative and alleviating the destructive.

There are many sobering analyses of the destructive consequences of automation, with some estimates that up to half of all jobs are at risk by 2030. PwC estimates that up to ten million jobs in the UK, 30 per cent of employment, could be at high risk by the early 2030s. Automation most clearly affects routine and repetitive manual tasks, and sectors such as transport, manufacturing, and wholesale are considered high risk, but other sectors are also under threat. The increasing use of automated processes such as machine-learning is affecting industries that include banking, insurance, and the legal profession. The machine learns from patterns in vast amounts of data and develops its own rules for how to interpret new information so it can solve problems and learn with very little human input. Its intelligence lies in being able to convert unstructured information into useful knowledge. Machine-learning is being used, for example, in insurance claims, automated tax filings, and in criminal justice cases. Many professionals—accountants, lawyers, consultants, doctors—are concerned about changes in their work.

Leading scientists and entrepreneurs, such as Stephen Hawking and Elon Musk, have been very concerned about the impact of AI, and especially that element of it that is autonomous, unsupervised by humans. Both have expressed fears for a world where AI becomes so powerful that people will no longer develop ideas as quickly as do machines. The question is how to encourage the positives of these innovations and ameliorate their negatives?

Machines that are capable of learning can help humans make better decisions. Demis Hassabis, a pioneer of AI, believes the technology will help society save the environment, cure disease, explore the universe, and also better understand ourselves as humans. Computers increased the speed and reduced the cost of calculation, and AI and machine-learning can increase the speed and reduce the cost of prediction and discovery. This is especially helpful where information is imperfect and piecemeal, and problems highly complex, such as in addressing climate change or curing Alzheimer's disease. There are opportunities to use machine-learning alongside human decision-making in circumstances where there is high unpredictability, and significant social interaction and personal discretion over tasks.

It is important that automation in future is used to augment rather than replace human judgement. AI and machine-learning can give doctors better information about patients, analysing results from scans and blood tests, for example, against vast populations of data so as to develop personalized diagnoses and care, but well-informed clinicians always need to be in control of decisions. Answers from machines should not be accepted unless they can explain how their conclusions were reached and can be assessed by knowledgeable people. Furthermore, as machine-learning captures the state of existing knowledge it reflects how things are, rather than how they could or ought to be. Users of these technologies need to understand its limitations, how insights and future ambitions could be constrained by biases in the existing data and reliance on what is currently known.

It remains crucial to overlay ethical considerations, empathy, simple human decency, and common sense on any machine-derived solutions to problems.

Human intuition and judgement has continually to be emphasized and encouraged. Humans are brilliantly capable of using the mind's eye to imagine things and connections impossible to see in the digital world. As Karl Marx said, the architect differs from the bee because humans see in their imagination before creating in reality. It is hard to imagine how AI could conceive Picasso's *Guernica*, write *Pride and Prejudice*, or compose *The Magic Flute*. As a US study of the impact of machine-learning and automation for the executive office of the president, says, AI 'still cannot replicate social or general intelligence, creativity, or human judgment', and that in the future 'employment requiring manual dexterity, creativity, social interactions and intelligence, and general knowledge will thrive'. A report written for the UK prime minister on the future of work concludes that automation and AI

> can only function as effective enablers of human productivity if their design takes full account of human aspirations, autonomy, behaviour and limitations. This is a real opportunity to ensure automation enhances the working experience rather than rendering it redundant.

Martin Ford's *Rise of the Robots* paints a disturbing picture of the impact of automation on work. Yet again there is a positive element. Robots need to be manufactured, requiring facilities, capital, and labour. Vast numbers of robots are used in assembly tasks in factories, improving quality and efficiency, and taking over many repetitive and exhausting jobs. Automated mining trucks and driverless farm vehicles, which are essentially robots, are enhancing productivity in mineral extraction and agriculture. Robots provide prostheses that respond to neurological signals, and help the elderly by providing them assistance with dignity. They can also be used for difficult and dangerous tasks. Robots are used in surgical operations because they are accurate, highly stable (no hand tremors), and are capable of bringing additional information to the surgeon's task, for example, through augmented reality where reality is supplemented by computer-generated content about a patient's condition. When it comes to safety, why would a human life be risked in fighting a fire, repairing operating gas pipelines, or disposing a bomb, if a robot could do the job?

The process of creative destruction is being played out in continuous streams of innovations. Extraordinary automation technologies are being developed that will affect everyday life. Drones to deliver parcels immediately following a purchase; direct language translation in real time; robots that clean and press clothes; 3D printing new materials—the list is endless. Some of these technologies will bring profound benefits, others will have a mixed record, many will fade into insignificance. Their diffusion and impact, their creativity and destructiveness, will result from the interaction of a huge range of economic, social, cultural, and political factors. Success will not only depend on their intrinsic costs and value to people, but on government regulations and changes in personal and group behaviour, and social attitudes. Returning to the driverless car, new regulations are needed to help answer the question, for example, of who is liable if the car crashes: the car maker, software provider, or the passenger? New behaviours will be needed if more cars are pooled, shared, and ordered on demand, and the status of owning a particular car, so important to many, is diminished.

Social acceptance of innovation is greater when it benefits society as a whole, rather than, say, just the business community. AI and machine-learning will advance further when its proponents move beyond promotion of its clever technologies and impacts on productivity. It will become more meaningful, and there will be greater confidence in it when, for example, people see its advantages in predicting, and hence helping avoid, problems that can affect everyone, such as disease, addiction, or traffic congestion. The blockchain, as a tool that stores information in the cloud and not in a centralized system overseen by a major company such as a bank, has been crucial for the development of digital money, such as Bitcoins, but at present is only significant to a minority of people. The attractiveness of blockchain more generally, however, is that it gives individuals access to their personal records in a way that they alone control. When it is used for keeping passwords or personal financial records or important documents in a completely safe and immediately accessible manner, the technology will become much more widely diffused. Among the most daunting challenge confronting society in the age of digital technology is the threat of cybercrime and cyberterrorism. Priority has to lie with developing the security-enhancing abilities of technologies such as blockchain and quantum computing.

Alleviating the destructive side of automation will also require imaginative new social initiatives. The disruption confronting whole industries and sectors will need policies for re-skilling, and perhaps re-locating, those workers affected. The reduction of entry-level jobs in fields such as accounting and law, which can be repetitive but offer useful training and introductions to the professions, will disproportionately affect young people. New approaches to training young people are needed, and serious consideration is required of a social wage, where people receive payments on the basis of being citizens, rather than employees. Education and training systems need to be re-adjusted towards life-long learning and to people starting businesses or working in professional freelance roles rather preparing them to work permanently for someone else. Greater social value has to be placed on the caring professions, on those with the empathy to assist the elderly and infirm. New taxation regimes will be needed to pay for these adjustments. These could range from Bill Gates's suggested tax on robots, to greater attention to inheritance tax to alleviate continuing inequalities. Furthermore, there is a pressing need for the development of the technology to be directed and influenced by a broad range of stakeholders, including the people who will use and be affected by it. This, most clearly, requires greater diversity among technologists in the field, especially requiring more women, bringing their creative and collaborative skills.

People

How will we deal personally with the way innovation is changing? Whether we work in the private or public sector, community groups, or as a member of the public, how can we be smarter in the way we develop and use innovation? What will the impact be of AI on the ways we live, work, play, and innovate: will it replace or augment our human capacities? Increased technological literacy will certainly improve our effectiveness in the massively connected world. But we shall also have to become more adept at encouraging creativity, dealing with change, communicating across boundaries, and putting ideas into practice. Intuition and judgement, tolerance and ethical responsibility, diversity of interests and cross-cultural sensitivities are needed. Our capacities to think about new ideas, play with them by tinkering, testing, and prototyping, and rehearsing, implementing or doing them, have to be balanced. Our scepticism and critical faculties should be attuned to questioning 'this is the way it is', and capacities accentuated for articulating 'this is the way we want it'. We shall demand the rewards that Edison's laboratory workers experienced—although perhaps without the exhausting hours or fear of the corpse reviver. Indeed, with the wealth created through our knowledge, we expect job satisfaction in enriching workplaces, conducive to diversity, which fit in with our lifestyles, family circumstances, and choices. We should ensure the inventors and innovators who contribute so much—the Stephanie Kwoleks of the world—are recognized in a way that sport stars and entertainers are appreciated today.

Innovation is a restless process that brings with it continual uncertainty about its success and failure. It can be threatening as well as rewarding. How well we respond to it depends upon how open-minded and cooperative we are, how prepared we are to accept risk, and give space for the unusual and work with others who are differently minded. This will be influenced by the culture of organizations and the quality of leaders who recognize that job security and toleration of failure are crucial to innovation, that no one holds all the answers, progress is collaborative, and reputation lies in modesty in claims and professionalism in delivery.

The outcomes of innovation are not always beneficial, and their consequences often cannot be foretold. Adding lead to petrol solved the problem of engine knocking, but it left a disastrous environmental legacy. Thalidomide reduced morning sickness in pregnant mothers, but it induced disabilities in their babies. The dangerous divorce between action and consequence was clearly seen in the global financial crisis of 2008/9 where financial innovations were introduced without any checks and balances or consideration for their implications. Concern for the consequences of innovations has to be paramount among those who seek to introduce them, and designers of new technologies have to prioritize the ways they can augment meaningful human work, rather than replacing it.

The massive amounts of data available on individuals, to other people, corporations, and the state, also increase the responsibilities of those designing and managing innovation. Innovation—in information use and other areas such as genetics—requires deep ethical considerations, highly visible and accountable practices, and alert and responsive regulations. Simulation, modelling, and virtualization technologies provide huge opportunities for improving innovation processes, but their responsible use depends on the skills and judgement of people immersed in the theory and craft of their professions and trades. Innovation requires people to be informed, vigilant, and responsible employees, customers, suppliers, collaborators, team members, and citizens. Andrew Grove, the founder of Intel, said that in our uncertain world only the paranoid survive, but it will be the discerning and informed, not the mistrusting and fearful, that will see us through. Immanuel Kant said that science is organized knowledge; wisdom is organized life. The future of innovation—where its benefits flow and costs are curtailed—lies in the wise organization of knowledge.

References

W. Abernathy and J. Utterback (1978). 'Patterns of Industrial Innovation', *Technology Review* 80(7): 40–7.

N. Baldwin, *Edison: Inventing the Century* (New York: Hyperion Books, 1995).

W. Baumol, *The Free-Market Innovation Machine: Analyzing the Growth Miracle of Capitalism* (Princeton, NJ: Princeton University Press, 2002).

M. Boden, *The Creative Mind: Myths and Mechanisms*, 2nd edn (London: Routledge, 2004).

E. Brynjolfsson, and A. McAfee, *The Race Against the Machine* (London: Digital Frontier Press, 2011).

T. Burns and G. Stalker, *The Management of Innovation* (London: Tavistock Publications, 1961).

H. Chesbrough, *Open Innovation: The New Imperative for Creating and Profiting from Technology* (Cambridge, MA: Harvard Business School Press, 2003).

C. M. Christensen, *The Innovator's Dilemma: When New Technologies Cause Great Firms to Fail* (Boston, MA: Harvard Business School Press, 1997).

L. Dahlander and D. Gann (2010), 'How Open is Innovation', *Research Policy* 39(6): 699–709.

A. Davies, D. Gann, and T. Douglas (2009), 'Innovation in Megaprojects: Systems Integration in Heathrow Terminal 5', *California Management Review* 51(2): 101–25.

M. Dodgson, D. Gann, S. MacAulay, and A. Davies (2015), 'Innovation Strategy in New Transportation Systems: The Case of Crossrail', *Transportation Research Part A: Policy and Practice* 77: 261–75.

M. Dodgson, D. Gann, and A. Salter (2006), 'The Role of Technology in the Shift Towards Open Innovation: The Case of Procter & Gamble', *R&D Management* 36(3): 333–46.

M. Dodgson, D. Gann, and A. Salter (2007), '"In Case of Fire, Please Use the Elevator": Simulation Technology and Organization in Fire Engineering', *Organization Science* 18(5): 849–64.

M. Dodgson, D. Gann, I. Wladawsky-Berger, N. Sultan, and G. George (2015), 'Managing Digital Money', *Academy of Management Journal* (Editor's Invited Article) 58(2): 325–33.

M. Dodgson and L. Xue (2009), 'Innovation in China', *Innovation: Management, Policy and Practice* 11(1): 2–6.

Executive Office of the President, *Artificial Intelligence, Automation, and the Economy* (Washington, DC: Executive Office of the President, 2016).

G. Fairtlough, *Creative Compartments: A Design for Future Organisation* (London: Adamantine Press, 1994).

M. Ford, *Rise of the Robots: Technology and the Threat of a Jobless Future* (London: Basic Books, 2015).

C. Freeman and C. Perez, 'Structural Crises of Adjustment: Business Cycles and Investment Behaviour', in G. Dosi, C. Freeman, R. Nelson, G. Silverberg, and L. Soete (eds), *Technical Change and Economic Theory* (London: Pinter, 1988).

L. Gerstner, *Who Says Elephants Can't Dance: Inside IBM's Historic Turnaround* (New York: Harper Business, 2002).

A. B. Hargadon, *How Breakthroughs Happen: The Surprising Truth about How Companies Innovate* (Cambridge, MA: Harvard Business School Press, 2003).

C. Helfat, S. Finkelstein, W. Mitchell, M. Peteraf, H. Singh, D. Teece, and S. Winter, *Dynamic Capabilities: Understanding Strategic Change in Organizations* (Malden, MA: Blackwell, 2007).

R. Henderson and K. B. Clark (1990), 'Architectural Innovation: The Reconfiguration of Existing Product Technologies and the Failure of Established Firms', *Administrative Science Quarterly* 35(1): 9–30.

C. Kerr, *The Uses of the University* (Cambridge, MA: Harvard University Press, 1963).

R. K. Lester, *The Productive Edge* (New York: W. W. Norton & Co., 1998).

B. A. Lundvall (ed.), *National Innovation Systems: Towards a Theory of Innovation and Interactive Learning* (London: Pinter, 1992).

F. Malerba, *Sectoral Systems of Innovation: Concepts, Issues and Analyses of Six Major Sectors in Europe* (Cambridge: Cambridge University Press, 2004).

K. Marx, *Capital*, vol. 1 (Harmondsworth: Pelican, 1981).

A. Millard, *Edison and the Business of Innovation* (Baltimore, MD: Johns Hopkins University Press, 1990).

R. Nelson and S. Winter, *An Evolutionary Theory of Economic Change* (Cambridge, MA: Belknap Press, 1982).

R. Nelson (ed.), *National Innovation Systems: A Comparative Analysis* (New York: Oxford University Press, 1993).

D. F. Noble, *Forces of Production: A Social History of Industrial Automation* (New York: Oxford University Press, 1986).

C. Paine, *Who Killed the Electric Car?* (Documentary film) (California: Papercut Films, 2006).

R. Parmar, I. Mackenzie, D. Cohn, and D. Gann (2014), 'The New Patterns of Innovation', *Harvard Business Review* 92(1–2): 86–95.

J. Quinn (2003), 'Interview with Stephanie Kwolek', *American Heritage of Invention and Technology* 18(3) <http://www.americanheritage.com>.

E. M. Rogers, *Diffusion of Innovations*, 4th edn (New York: The Free Press, 1995).

R. Rothwell, C. Freeman, A. Horley, V. Jervis, Z. Robertson, and J. Townsend (1974), 'SAPPHO Updated—Project SAPPHO, Phase II', *Research Policy* 3: 258–91.

Royal Society, Hidden Wealth: *The Contribution of Science to Service Innovation* (London: Royal Society, 2009).

Royal Society of Arts, *Good Work: The Taylor Review of Modern Working Practices* (London: Royal Society of Arts, 2017).

K. Sabbagh, *Twenty-First-Century Jet: The Making and Marketing of the Boeing 777* (New York: Scribner, 1996).

J. A. Schumpeter, *The Theory of Economic Development: An Inquiry into Profits, Capital, Credit, Interest and the Business Cycle* (Cambridge, MA: Harvard University Press, 1934).

J. A. Schumpeter, *Capitalism, Socialism and Democracy* (London: George Allen & Unwin, 1942).

S. Smiles, *Josiah Wedgwood: His Personal History* (London: Read Books, 1894).

A. Smith, *An Inquiry into the Nature and Causes of the Wealth of Nations* (London: Ward, Lock and Tyler, 1812).

D. Stokes, *Pasteur's Quadrant: Basic Science and Technological Innovation* (Washington, DC: Brookings Institution Press, 1997).

D. J. Teece (1986), 'Profiting from Technological Innovation: Implications for Integration, Collaboration, Licensing and Public Policy', *Research Policy* 15(6): 285–305.

J. Uglow, *The Lunar Men: Five Friends Whose Curiosity Changed the World* (New York: Farrar, Straus and Giroux, 2002).

J. M. Utterback, *Mastering the Dynamics of Innovation: How Companies Can Seize Opportunities in the Face of Technological Change* (Boston, MA: Harvard Business School Press, 1994).

J. Womack, D. Jones, and D. Roos, *The Machine that Changed the World: The Story of Lean Production* (New York: Harper, 1991).

J. Woodward, *Industrial Organization: Theory and Practice* (London: Oxford University Press, 1965).

Further reading

On Josiah Wedgwood:

M. Dodgson (2011), 'Exploring New Combinations in Innovation
and Entrepreneurship: Social Networks, Schumpeter, and the
Case of Josiah Wedgwood (1730–1795)', *Industrial and Corporate
Change* 20(4): 1119–51.

On Joseph Schumpeter:

T. McGraw, *Prophet of Innovation: Joseph Schumpeter and Creative
Destruction* (Cambridge, MA: Harvard University Press, 2007).

On the innovation process, and the ways it is organized, managed,
and changing:

M. Dodgson, D. Gann, and A. Salter, *Think, Play, Do: Technology,
Innovation and Organization* (Oxford: Oxford University Press,
2005).

M. Dodgson, D. Gann, and A. Salter, *The Management of
Technological Innovation: Strategy and Practice* (Oxford: Oxford
University Press, 2008).

M. Dodgson, D. Gann, and N. Phillips (eds), *The Oxford Handbook of
Innovation Management* (Oxford: Oxford University Press, 2014).

M. Dodgson (ed.), *Innovation Management: Critical Perspectives on
Business and Management*. Volume 1: *Foundations*; Volume 2:
Concepts and Frameworks; Volume 3: *Important Empirical
Studies*; Volume 4: *Current and Emerging Themes* (London:
Routledge, 2016).

On the economics of innovation:

J. Fagerberg, D. Mowery, and R. Nelson (eds), *The Oxford Handbook
of Innovation* (Oxford: Oxford University Press, 2005).

J. Foster and J. S. Metcalfe (eds), *Frontiers of Evolutionary Economics* (Cheltenham: Edward Elgar, 2003).

On the history of innovation:

D. Edgerton, *Shock of the Old: Technology and Global History Since 1900* (London: Profile Books, 2006).

N. Rosenberg, *Inside the Black Box: Technology and Economics* (Cambridge: Cambridge University Press, 1982).

On innovation strategies:

M. Schilling, *Strategic Management of Technological Innovation* (New York: McGraw-Hill/Irwin, 2005).

On entrepreneurship:

G. George and A. Bock, *Inventing Entrepreneurs: Technology Innovators and their Entrepreneurial Journey* (London: Prentice Hall, 2009).

On the impact of artificial intelligence:

PwC (n.d.), 'Sizing the Prize', <www.pwc.com/gx/en/issues/data-and-analytics/publications/artificial-intelligence-study.html> (last accessed 10 August 2017).

Data on international R&D and innovation performance:

National Science Foundation, 'Science and Engineering Statistics' <www.nsf.gov/statistics>.

Organisation for Economic Co-operation and Development (OECD), 'Science, Technology and Patents', Statistics Portal: <www.oecd.org>.

Index

Innovation

FASHION
A Very Short Introduction
Rebecca Arnold

Fashion is a dynamic global industry that plays an important role in the economic, political, cultural, and social lives of an international audience. It spans high art and popular culture, and plays a significant role in material and visual culture. This book introduces fashion's myriad influences and manifestations. Fashion is explored as a creative force, a business, and a means of communication. From Karl Lagerfeld's creative reinventions of Chanel's iconic style to the multicultural reference points of Indian designer Manish Arora, from the spectacular fashion shows held in nineteenth century department stores to the mix-and-match styles of Japanese youth, the book examines the ways that fashion both reflects and shapes contemporary culture.

'Her fascinating little book makes a good framework for independent study and has a very useful bibliography.'

Philippa Stockley, Times Literary Supplement

ORGANIZATIONS
A Very Short Introduction
Mary Jo Hatch

This *Very Short Introductions* addresses all of these questions and considers many more. Mary Jo Hatch introduces the concept of organizations by presenting definitions and ideas drawn from the a variety of subject areas including the physical sciences, economics, sociology, psychology, anthropology, literature, and the visual and performing arts. Drawing on examples from prehistory and everyday life, from the animal kingdom as well as from business, government, and other formal organizations, Hatch provides a lively and thought provoking introduction to the process of organization.

www.oup.com/vsi

ADVERTISING
A Very Short Introduction
Winston Fletcher

The book contains a short history of advertising and an
explanation of how the industry works, and how each of the
parties (the advertisers , the media and the agencies) are
involved. It considers the extensive spectrum of advertisers
and their individual needs. It also looks at the financial side of
advertising and asks how advertisers know if they have been
successful, or whether the money they have spent has in fact
been wasted. Fletcher concludes with a discussion about the
controversial and unacceptable areas of advertising such as
advertising products to children and advertising products such
as cigarettes and alcohol. He also discusses the benefits of
advertising and what the future may hold for the industry.

ECONOMICS
A Very Short Introduction
Partha Dasgupta

Economics has the capacity to offer us deep insights into some of the most formidable problems of life, and offer solutions to them too. Combining a global approach with examples from everyday life, Partha Dasgupta describes the lives of two children who live very different lives in different parts of the world: in the Mid-West USA and in Ethiopia. He compares the obstacles facing them, and the processes that shape their lives, their families, and their futures. He shows how economics uncovers these processes, finds explanations for them, and how it forms policies and solutions.

'An excellent introduction . . . presents mathematical and statistical findings in straightforward prose.'

Financial Times

GENIUS
A Very Short Introduction
Andrew Robinson

Genius is highly individual and unique, of course, yet it shares a compelling, inevitable quality for professionals and the general public alike. Darwin's ideas are still required reading for every working biologist; they continue to generate fresh thinking and experiments around the world. So do Einstein's theories among physicists. Shakespeare's plays and Mozart's melodies and harmonies continue to move people in languages and cultures far removed from their native England and Austria. Contemporary 'geniuses' may come and go, but the idea of genius will not let go of us. Genius is the name we give to a quality of work that transcends fashion, celebrity, fame, and reputation: the opposite of a period piece. Somehow, genius abolishes both the time and the place of its origin.